IMAGES
of Rail

ILLINOIS MIDLAND RAILWAY

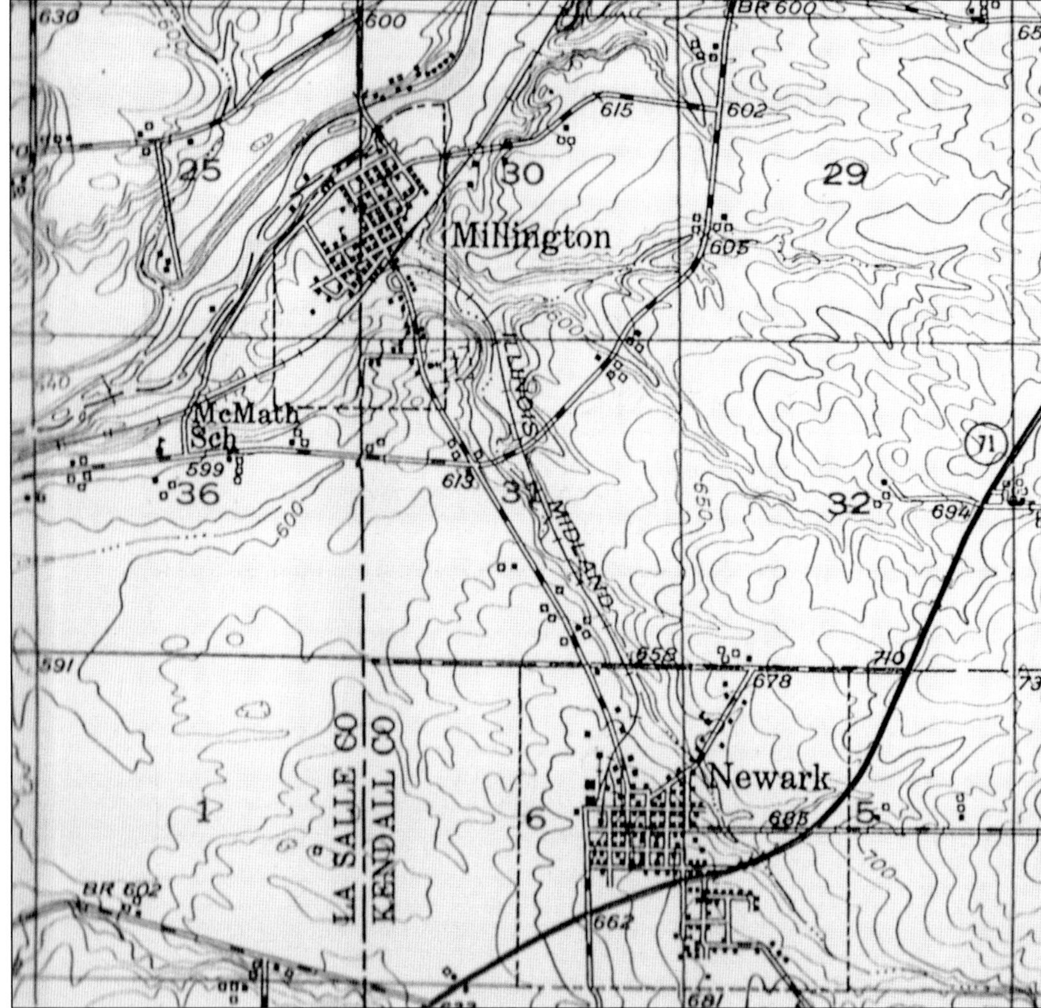

A map of the Illinois Midland Railway does not take up much room in a book—or anywhere else. What a map cannot show is the picturesque countryside it covered—the fields, the pastures, the creek (which needed five bridges to cross) where water was drawn for use in the locomotive's boiler, the slight but slippery hill coming into Newark, the grove of oak trees by Millington where camp meetings were once held, and the handful of men who successfully ran the operation for 53 years. (Kendall County Historical Society.)

On the Cover: In 1914, there was something to celebrate when the Illinois Midland Railway connected two small towns, Millington and Newark. Even though the new railroad was less than two miles in length, it never went any farther. F.A. Schafer was the engineer sitting in the cab of a borrowed locomotive, while a homemade tool car stood on an adjacent track, and the rest of the crew might have been wondering where Samuel Graham Durant, the railroad's founder and self-proclaimed president, was on that day. He was headed to prison. The Midland, as it was called, was off to a shaky financial start but would survive for 53 more years as the so-called "Shortest Railroad in the World." (Millington, Illinois, Historical Society Museum.)

IMAGES
of Rail

Illinois Midland Railway

Jeff Kehoe

ISBN 978-1-4671-0775-4

Published by Arcadia Publishing
Charleston, South Carolina

Printed in the United States of America

Library of Congress Control Number: 2021949620

For all general information, please contact Arcadia Publishing:
Telephone 843-853-2070
Fax 843-853-0044
E-mail sales@arcadiapublishing.com
For customer service and orders:
Toll-Free 1-888-313-2665

Visit us on the Internet at www.arcadiapublishing.com

This book is dedicated to Beverly Shields Casey of Millington, Lowell Mathre of Newark, and all the men and women of the surrounding areas who remember the Illinois Midland—it was their railroad!

Contents

Acknowledgments

This project started as a family effort. A suggestion to write a book about this short line railroad came from Jenn Kehoe, my daughter-in-law, then my sons Tom and Brian got involved with research and proofreading, plus my grandson, Michael Kehoe, was my tech support. Thanks go to them and Bev Casey, director of the Millington, Illinois, Historical Society Museum; Lowell and Jeff Mathre of Fern Dell Museum in Newark; Marilyn Thompson and the board of directors at Fern Dell Museum; Roger Matile of Little White Schoolhouse Museum in Oswego; Thomas Whitt of Burlington Route Historical Society; Lisa Wolancevich of Kendall County Historical Society; Jimmy Judd; Wayne DeMunn; Doug Holley; Dick West; Janet Luther Blue; Vic Stott; and Joe Sorensen. The old photographs in this book from my collection that I have wanted to share are credited to "Author's collection" and were given to me by my grandfather A.E. Shaw and a friend, Robert Janz. Photographs credited "Author's photograph" are pictures I personally took over a span of more than 30 years.

Introduction

What this railroad is and is not needs clarification at the start of this story. Take its name, for instance—the Illinois Midland Railway. No book has been written about it until now, but there was a railroad in this state with a similar name—the Chicago and Illinois Midland (C&IM) Railroad. That railroad had Chicago in its name, although it never reached Chicago. At over 120 miles in length and a route south from Peoria through Springfield to Taylorville, the C&IM was the same length that the Illinois Midland was when it was on the drawing board, so to speak, but the Illinois Midland discussed in this book was officially just 1.962 miles long—they are *not* the same railroad.

At nearly two miles in length, the Illinois Midland (or just "the Midland") connected the towns of Newark and Millington, Illinois, and terminated at the depot of the Chicago, Burlington & Quincy (CB&Q, or Burlington) Railroad in Millington about a quarter-mile south of the Fox River. The Burlington Railroad was much larger and more established, having originated in 1850 at nearby Aurora. Much of the Midland's history is linked to the Burlington and Kendall County, as well as other railroads in the region that had a role in its legacy.

The Midland did not become the successor to the C&IM; that successor is known as the Illinois and Midland (IMRR), which was started in 1996 as a subsidiary of Genesee and Wyoming (G&W) Rail Link running from Pekin to Taylorville using some of the same trackage as the C&IM. The IMRR has nothing to do with what happened in Newark and Millington.

The Midland was not always the shortest railroad in the nation, but in the 1950s and 1960s, it was listed in the *Guinness Book of Records* as such. It was recognized for decades as the shortest standard-gauge railroad in the state, and although that might be a matter of debate today, many railfans maintain that the Midland, of all known rail lines, still holds the record as the shortest in Illinois.

This was not a railroad that operated alone—no railroad does. Instead, the connections any railroad makes with other lines become important to its survival. Freight, the major source of income for most railways, seldom travels along a single line from its point of origin to its destination. Other lines get involved and cars are interchanged, and with the Illinois Midland, the Burlington proved to be a good connection for over 50 years. In a way, the Burlington was a benevolent big brother, and the nearby Milwaukee Road was a good neighbor. A look at what was going on with both of these railroads helps explain the scene of the times.

The genesis of the Midland began with an idea that came from a fellow who called himself a railroad promoter, executive, engineer, financier, and stockbroker, among other titles. There is no evidence he had formal training in any of these vocations, however. His name was Samuel Graham Durant, although his birth surname was Bondurant, and his early years are a bit of a mystery. He claimed to have been born in New York on June 24, 1882, but other information places him in Jefferson County, Indiana, a rural area in the southern part of the state that had been home to his family for years.

What is known about Durant is that he was a smooth-talking con artist, and the way any con artist operates is to first gain the confidence of the people they are about to swindle. The residents of Newark and Millington fell for his scheme because he told them what they wanted to hear—that his railroad would bring prosperity and prominence to their communities. His storyline was well rehearsed, as Durant had used it in other small communities across America for a decade before he arrived in Kendall County, Illinois. Basically, he sold shares of stock to local folks to help cover the costs of building a railroad that they had been convinced was something they needed. Durant would then disappear with the money, leaving the locals with worthless shares of stock. This book includes a rare photograph of Durant.

The first time Durant appears to be mentioned in any newspaper was in early 1906, when the 24-year-old Bondurant (before losing the "Bon" prefix) arrived in Florida. He chose the little town of Arcadia in DeSoto County near Sarasota as the target for his first scam, bluffing his way into making people believe he was a Philadelphia financier trying to build a marvelous new railroad. He gained the confidence of some of its prominent citizens, pitching a proposal that involved him spearheading the construction of a railroad that would run from the East Coast to the West Coast, and it would run through Arcadia, turning their town into a great city. Even more amazing was his claim that the entire route would be electrified, which was a novel idea for 1906. Once the money had been collected, Durant disappeared, leaving angry citizens with worthless stock and no railroad!

Durant perpetrated his scheme again in Texas, but his undoing came over the fact that he was using the mail to distribute his worthless stock, which meant that he was committing mail fraud. On September 24, 1907, Durant was arrested for mail fraud—a federal offense. He posted a $2,500 bond in Dallas before fleeing to Canada.

A century ago, Canada might have been a good place to hide, but Durant got into trouble there too. After he fled from Canada, authorities induced Durant to come across the border into Plattsburg, New York, where, after posting $3,500 in bail money, his case was suspiciously never called to trial. He then headed across the Midwest to Nebraska, Iowa, and Illinois, always managing to stay a step ahead of the law.

Durant was in Illinois in 1913 with a plan to build a railroad between Rockford and Kankakee; it was the Illinois Midland, and this time, he did produce a railroad—although it was built with substandard materials and did not span the promised 120 miles connecting Rockford and Kankakee, instead traversing less than 2 miles between Newark and Millington. Despite these limitations, on February 14, 1914, three hundred people in the two little towns turned out for the big opening day of the Midland. Durant, however, was not present for the festivities, because the long arm of the law had finally caught up with him, and he was headed for prison!

News of Durant's unexpected departure hit Newark and Millington like a bomb as both towns learned they had fallen victim to another one of his schemes. Newark and Millington have always had similarities and differences. Both were founded around the same time—Newark in 1838 and Millington in 1835. Pioneers like George Hollenback and Samuel Jackson had arrived a few years earlier. The two towns and others like them in Kendall County had the stories of their earliest days detailed in a book called *The History of Kendall County Illinois*, written by Rev. E.W. Hicks and published in 1877. Both towns began under different names than they are known by today; Millington was Milford, and Newark was Georgetown. Both share a common cemetery—the Millington-Newark Cemetery on Fox River Drive. Both were stops along the Underground Railroad, which had nothing to do with a real railroad but remains an interesting chapter in the area's history from the years before the Civil War.

This book will explain how both Newark and Millington had the potential to become major cities 100 years ago, although a combination of circumstances resulted in both remaining small towns to this day. Today, Newark's population is around 1,500, while Millington's is a little over 500. Railroads played an important part in what happened to the towns in those 100 years.

George B. Hollenback, recognized as Newark's first settler (hence its original name of Georgetown), ran a trading post. In time, the settlement became a farming community, and that enterprise continues today. Located two miles south of the Fox River and situated on higher ground, Newark had the advantage of being a popular stop on the stagecoach route from Chicago to Ottawa and points south and west. Several hotels sprang up, and they were often filled to capacity with travelers. Soon, other businesses in town were booming.

As railroads began replacing stagecoach lines across Illinois around the time of the Civil War, it became apparent that to continue its growth, Newark desperately needed to get rail service. As far back as 1858, a railroad had been chartered to run from Joliet to Mendota, passing through Newark. Preliminary grading and filling for its roadbed began at a cost of thousands of dollars, but nothing was completed. The outbreak of the Civil War drew attention elsewhere, leaving a half-dozen principal citizens in the area out of luck. Later attempts to build a line through Newark continued to fail until 1871, when a railroad was finally completed—the Ottawa, Oswego & Fox River Valley (OO&FRV)—but it went through Millington, not Newark!

Millington's history begins with the arrival of Samuel Jackson and George Markley, who bought land along the Fox River. Jackson, originally from south of Pittsburgh, met Markley on a steamboat headed west, and both men decided to settle along the Fox River in LaSalle County near the Kendall County line. Each man bought 300 acres on opposite sides of the river, but it was Jackson's land on the south side of the river that would develop into Millington. It was there that the men built a double log cabin—two standard-sized cabins spaced about 10 feet apart and joined by a common roof. They began damming the river above a small island and built a sawmill early in 1835. Samuel Jackson's brother, Jesse, arrived for a visit and decided to stay. According to local history, Markley was "having pretensions to the law," got himself elected as justice of the peace, then moved away. Jesse Jackson bought him out, and a year later, the rest of his family joined him, including 11-year-old Joseph Jackson. It was Joseph who eventually did the most for the town, according to Reverend Hicks's history of Kendall County. In his 40s, Joseph helped bring the OO&FRV Railroad from imagination into reality, and with it came business opportunities for Millington.

As more people arrived, the town grew and prospered, with farming becoming its major business. The discovery of silica sand nearby brought another industry—mining. Millington had churches, shops, hotels, and factories; one interesting factory produced buttons by harvesting clams out of the Fox River and drilling into their shells.

However, problems can arise when a settlement is started alongside a river. History is full of stories about towns founded next to rivers, but what happens when the river floods? Such was the case with Millington and the Fox River. An early flood dating to 1868 caused damage that swept away the town's wooden bridge and dam. This was not the first nor the last time that sort of thing would happen. No railroad would want to build tracks too close to an unruly river, and that important fact was not lost on the construction company that set out to build along the Fox.

An independent railroad (with cheaper freight rates), the OO&FRV, was chartered, and it would run as far north as Aurora, pass somewhere near Newark or Millington, then extend as far south as Streator. Joseph Jackson had been an active member of the OO&FRV Railroad board since 1866, but when construction began in 1870, it remained to be seen if the line would come to Newark or Millington.

Millington was prosperous but prone to flooding. Newark was also prosperous but uphill, where flooding did not occur, although the next town downriver that the railroad would reach would be Sheridan, which was a few miles out of the way from Newark. Tracks into Millington could be built a quarter-mile uphill from the river, making them unlikely to flood, but the extra uphill and out-of-the-way miles to Newark could be avoided if Millington was on the chosen route. It also did not hurt that Millington's Joseph Jackson was on the railroad's board of directors.

Millington got the railroad, and the residents of Newark were furious!

Joseph Jackson "did more to promote the enterprise than any other man on the line," according to Reverend Hicks's book. There was disappointment among some of the townspeople when the Burlington got control of the new 57-mile railroad. It may have not been a bad thing, though, since under the Burlington's tutelage, the line prospered and is still in business today; it is now part of Illinois RailNet, not Burlington Northern Santa Fe (BNSF).

Newark's second chance at getting a railroad came when Durant arrived in 1913 with plans to build the Illinois Midland, but the unsuspecting citizens of Newark and the surrounding area had no way of knowing that he was using a phony stock scam to take their money. Durant's ideas for which directions the Midland would run were not very original; in fact, a similar railroad had been built a decade earlier—the Chicago, Milwaukee & Gary (CM&G, or "the Gary Line") Railway, which would become a subsidiary of the Chicago, Milwaukee, St. Paul & Pacific (CMStP&P or "the Milwaukee Road") Railroad in 1922. At just under 130 miles in length, the Gary Line connected the up-and-coming town of Rockford (to the northwest) across the prairie in a crescent-shaped curve with the area near Kankakee where surface mines in Ogle and Kankakee Counties produced cheap coal. Durant's imagined railroad was a 120-mile line that would start by connecting Newark and Millington, then extend north from Millington to Rockford and south from Newark to Kankakee. If this sounds familiar, it was; although it is impossible to say that Durant directly copied the Gary Line's concept, anybody in tune with the rail scene in Illinois at that time would have been familiar with the Gary Line. Durant's selling points for the Illinois Midland were cheaper coal arriving by the carload at a time when most homes and businesses used coal for heat and the economical transportation of agricultural products like grain, corn, and soybeans to benefit local farmers. With the addition of some coaches, passengers might even be able to travel from town to town. It sounded like a winner, so local folks could not be blamed for using their hard-earned money to buy into Durant's well-rehearsed plot.

Contrary to popular belief, Durant did not steal all the money and slip out of town late one night—he got the lion's share of the money legally, and this book will explain how he did it. Locals sprang into action, forming a kind of community enterprise to keep the railroad running and get a grain elevator built. The Midland succeeded despite its tumultuous beginning.

For most of its life, the Illinois Midland was a one-man operation, meaning that one man did every job on the railroad. Some of its engineers tell their stories in chapters three and four, as the quaint little railroad became well known. It is also intriguing to examine what the Midland might have been if it was built as planned and, in comparison, what it was like to work on a major road of those times and what equipment they used.

This railroad's ending in 1967 is not a happy one, but many people in the area where it once ran still fondly remember the Midland. It will always be their railroad.

One

The Promoter Arrives then Quickly Departs

A man financed a railroad by selling phony stocks, and when he left with the people's money, they wondered how he had done it.

This man's name was Samuel Graham Durant, a traveling con artist who had crossed the United States and parts of Canada with his fraudulent schemes, stealing the money of unsuspecting people via his railroad stock scams. In 1913, he came to Newark and Millington with a plan to build a 120-mile railroad that would put Newark "on the map," and make Millington into "the next Chicago" by building a line he called the Illinois Midland. All he needed to get started was to have them sign on to buy $25,000 in railroad stock.

Were there any takers in the two towns? As it turns out, there were. Durant knew what to say and how to say it, gaining people's confidence as well as their money. The two little towns, he said, would find themselves in the middle of a rail line connecting the cities of Rockford (to the northwest) with Kankakee (to the southeast) that was sure to become a success. Soon, the talk around town was that over $30,000 was on deposit in the Newark Farmers State Bank.

As president of the yet-unbuilt railroad with its prestigious Chicago headquarters (in reality, it was an apartment in Chicago's Hyde Park neighborhood), Durant voted himself 226 shares of his own stock at $100 per share. Work began on the first part of the line between Millington and Newark, and the contractors he hired were from a Chicago firm, the Monroe & Connors Construction Company, which was working in cahoots with Durant.

The smooth-talking Durant managed to get a railroad built between Newark and Millington, but it consisted of under two miles of substandard roadbed and rickety rails reminiscent of narrow-gauge routes, all spiked using secondhand ties and tracks. On February 14, 1914, a grand opening of sorts was held, and although the vast majority of the Illinois Midland Railway was still on the drawing board, local citizens were certain this was the start of something big and turned out in force. Newspapers like the *Aurora Beacon-News* reported on the gala event—it had a brass band, flag-waving, speechmaking, and a crowd of 300 on hand as free train rides were given using a borrowed locomotive with a couple of coaches. The newspapers speculated that the self-proclaimed president of the railroad—Durant—was absent due to health concerns. In truth, he was he was headed to prison!

When news of what really happened to Durant broke about a month later, it hit like a bomb. He was headed for federal prison in Atlanta, Georgia, because he had been convicted of fraud. His sentence was 15 months, but it had nothing to do with the Illinois Midland Railway. Rather, it was related to the role he had played in a complex stock scheme with a handful of other accomplices in Rochester, New York, in 1910; it was called the American Redemption Company and had nothing to do with railroads. According to news stories, a group of 14 men had swindled 500 people out of as much as $1 million. Whether those numbers are an exaggeration remains uncertain, but Durant got the stiffest sentence of all of them.

Townspeople in Newark and Millington, recovering from the shocking news, realized they would have to do something to rescue the unfinished railroad, and they decided to pool their money, forming a syndicate to get the railyard and grain elevator built in Newark. As the co-op was starting to unfold, someone from the past resurfaced.

After serving less than half his sentence, Durant was released in late July 1914 and returned to Newark of all places. What he wanted was to get his hands on whatever money was languishing in the Farmers State Bank of Newark, which was held in escrow and reputed to be $30,000.

However, no such amount of money existed.

In reality, most of the money existed only on paper, and that would be like trying to build a railroad financed by IOUs with no true monetary value. What cash there was amounted to around $12,000, and Durant claimed he was owed $10,000 of this for his "fees" that needed to be paid for the two miles of the railroad that had been completed. Indeed, he had put such an agreement in writing from the start of the Midland, so he began threatening to sue everyone who got in his way. Could a settlement be reached?

In fact, a settlement was reached, but not before the farmers' syndicate hired lawyers, and the whole thing ended up in court. There, Durant was represented by Aurora attorneys Peffers & Wing, one of the top law firms in the area. The lawyers for the syndicate argued that Durant had not finished building the railroad to Rockford or Kankakee as promised, but Durant's lawyers pointed to the fine print in the contract and countered that those two cities were merely points of destination made to illustrate where the Midland might end up, so if it only went as far as the two miles between Newark and Millington, Durant was still owed his fees. Furthermore, the Illinois Midland had never been sanctioned by the Illinois Public Utilities Commission to sell stock, so it was worthless. Durant was about to win.

The case went before Judge Susser in Aurora, but before he could render his decision, a settlement was reached. Durant would get almost $10,000 for his services, which he gladly accepted before vanishing from town, never to be seen in this part of the country again.

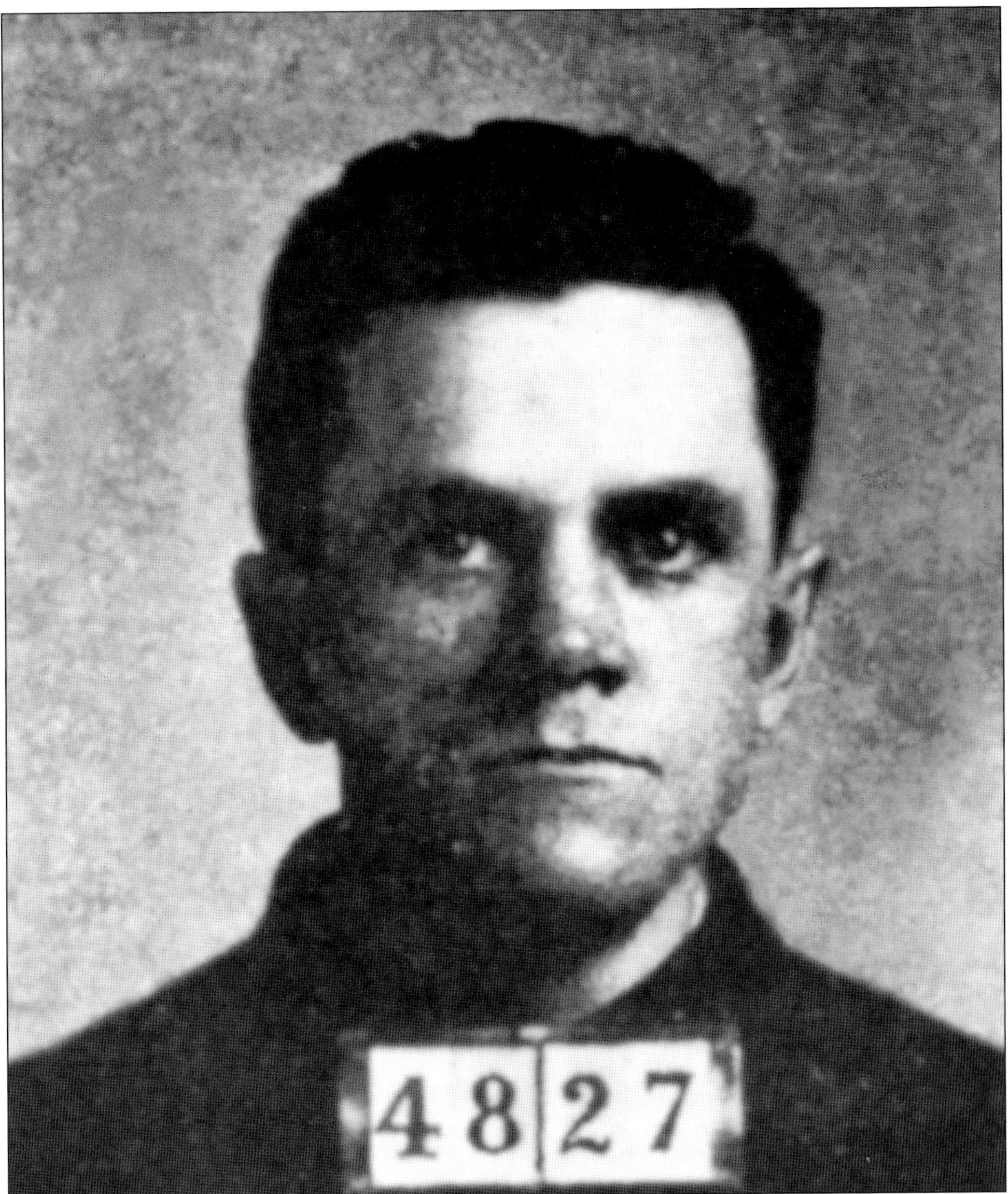

Mug shots are notoriously uncomplimentary, and this one is no exception—it features S.G. Durant, also known as Samuel Graham Bondurant (as well as a variety of other aliases). In a word, his crime was fraud. He perpetrated it in New York, Florida, Texas, Nebraska, and Iowa, then hid from the authorities for a time in Canada before coming to Newark, Illinois, in 1913 with a smooth sales pitch for building a railroad. In the end, he skipped town with the hard-earned money from citizens in Newark and Millington, leaving them with shares of worthless stock and two miles of rickety track. Very few photographs of Durant are known to exist, but this one from 1914 shows how he would have looked then—five feet, three inches tall, weighing 120 pounds, with brown hair and eyes. (Millington, Illinois, Historical Society Museum and Bev Casey.)

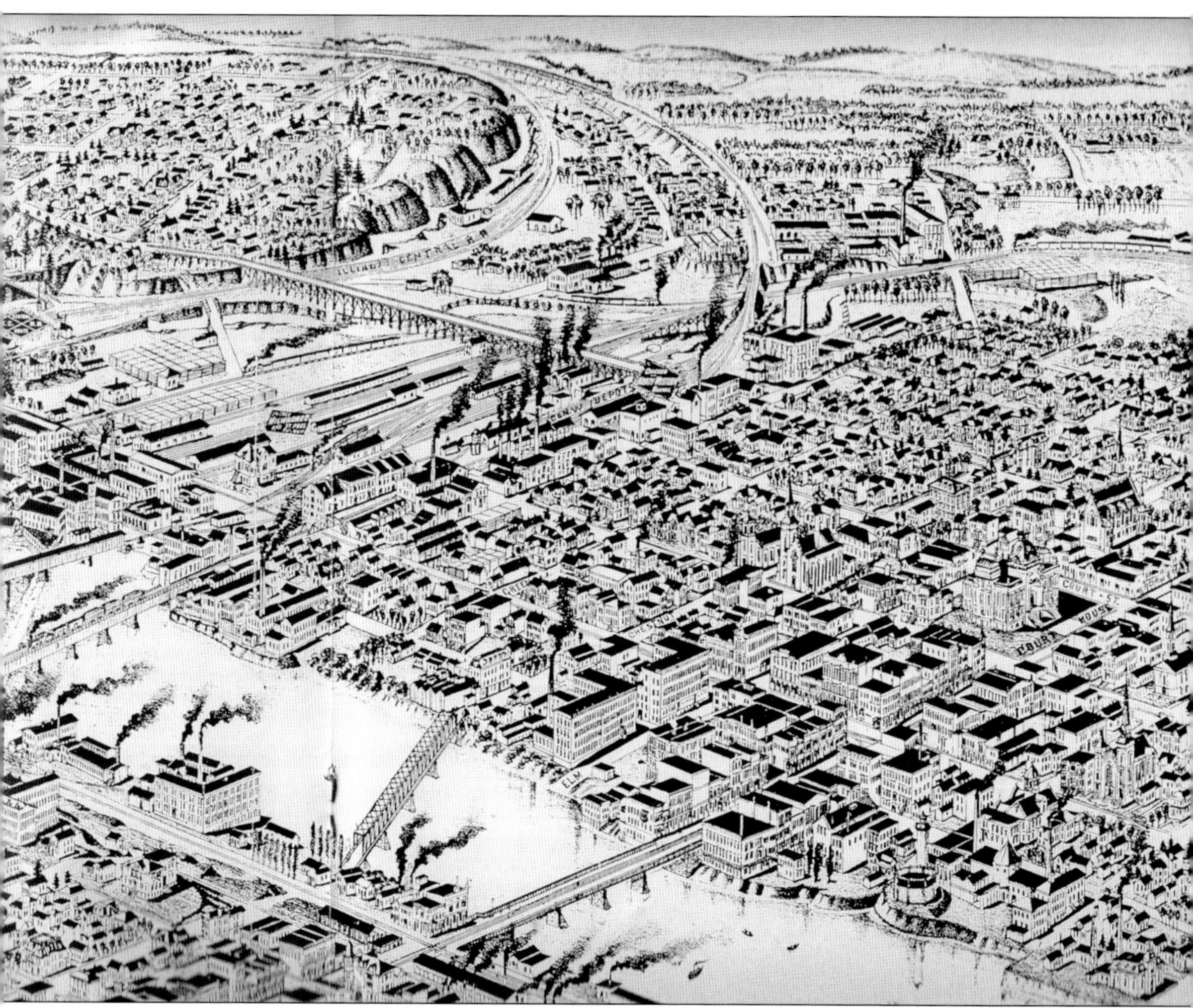

The city of Rockford, shown here in an artist's bird's-eye view from 1891, was the northernmost point of destination for the proposed Illinois Midland Railway. As a growing industrial city, it was second only to Chicago in importance around this time. The problem was that Rockford was already being served by three major railroads: the Milwaukee Road, Chicago & Northwestern, and Illinois Central. Samuel Graham Durant may have somehow believed his new rail would connect the Illinois Central at Rockford with another Illinois Central line that ran through Kankakee some 120 miles southeast. Such a connection—bypassing congested rail traffic in Chicago—would bring business and prosperity to both Newark and Millington, the two small towns anxious for a railroad of their own and situated in the middle of the action. (Newberry Library, Chicago.)

This share of stock in the Illinois Midland gives the appearance of legitimacy. At $100 apiece, they were expected to raise $25,000 toward building a 120-mile railroad from Rockford to Kankakee, but the Midland got only as far as about 2 miles. Close examination of this particular certificate reveals that S.G. Durant had voted himself 226 shares (which would equal over $500,000 dollars in 2021 money), but the shares were ultimately worthless. (Fern Dell Museum.)

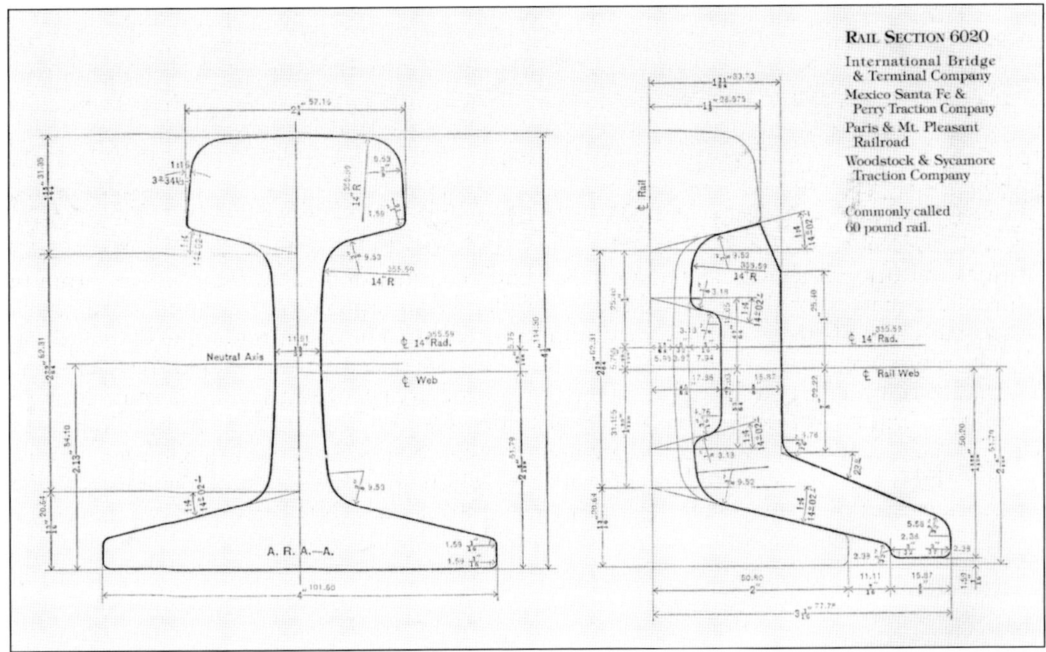

The rails that were used in building the Midland were known as lightweight, as shown in this 1926 diagram; lightweight rails were usually laid for interurban and trolley lines, not main line standard-gauge railways, which the Illinois Midland was supposed to be. Savings of a few dollars per yard easily translated into tens of thousands of dollars when figured into 100 miles or more of track being laid. (Author's collection.)

Train time at the depot could be exciting, offering a connection with the outside world. Millington already had a railroad—the Burlington's Fox River Line—that could take passengers the 57 miles from Aurora to Streator. What if a railroad like the Midland could connect to Newark and Kankakee or Rockford? This would make it possible for people to travel from one side of the state of Illinois to the other. (Millington, Illinois, Historical Society Museum.)

A good look at the borrowed equipment ready for opening day (and the cover of this book) on the Illinois Midland Railway shows passengers were expected to come out for a free ride from Millington to Newark. Although the new railroad was less than two miles in length, it was something to celebrate. A closer look reveals that the track was inferior in quality and construction. Not only was it too light to support heavy rolling stock, it was also bought secondhand, with even the ties being previously used. Note the lack of adequate ballast, which was needed for track stability and to provide drainage. Still, it was a much-anticipated accomplishment, so the townspeople were happy—at least for a while. (Millington, Illinois, Historical Society Museum.)

On February 14, 1914, opening day for the Illinois Midland Railway featured an appropriately numbered locomotive No. 1 and free rides for excited locals. A crowd of 300 enjoyed a brass band, speeches, and flag-waving, as reported by the *Aurora Beacon-News*. Rail was an important part of transportation, and now it connected the towns of Millington and Newark. (Millington, Illinois, Historical Society Museum.)

A less formal image of the Midland's 1914 opening day shows a good crowd. Perhaps it did not matter that it was a cold and snowy day or that the equipment was borrowed, but news had not yet reached them that the president and founder of the railroad was on his way to prison—although his crime had nothing to do with the Midland, S.G. Durant had been convicted of fraud and sentenced to 15 months in federal prison. (Millington, Illinois, Historical Society Museum.)

What appears to be just a simple picture that someone took on February 14, 1914, is an important record of what happened when the Midland celebrated its opening day. With people climbing all over the train, this may have been the moment just after it reached Newark. The arrival of the Midland was a newsworthy event, and the promise of a better future seemed to be just down the road. (Millington, Illinois, Historical Society Museum.)

What may be the only interior photograph taken on the Midland's opening day in February 1914 shows a carload of excited passengers enjoying their first ride on what was now their town's railroad. Several trips were reportedly made on that wintry February day, and although turbulent financial times awaited the little railroad, nothing could dampen the high spirits of the day. (Millington, Illinois, Historical Society Museum.)

A closer look at Illinois Midland locomotive No. 1 shows it to be an old-fashioned 4-4-0 wheel arrangement dating back to the time of the Civil War. Engines like this were already outdated, but many hung around for years in secondary service. Although it was a bit heavy for the Midland's rails, these so-called American-type locomotives were versatile and did well on uneven tracks like those that made up the Midland's short line. (Millington, Illinois, Historical Society Museum.)

This may be the same locomotive (or one nearly like it) from the previous image in use years earlier on the Milwaukee Road; at that time, it was No. 410. Built in 1871, the locomotive would have seen well over 40 years of service before being borrowed/leased to the Midland in 1914. What is known for certain is that Midland No. 1 was a former Milwaukee Road engine. (Milwaukee Road Historical Association.)

Two

How Not to Build a Railroad

Much of the work on the Illinois Midland was inadequate, but why did that happen, and what did other railroads do that was better?

In the latter half of 1913, Samuel Durant, self-appointed president of the Illinois Midland, began a rather feeble and shortsighted attempt at building a railroad. Since the money he had collected from his stock scheme came from people living in the surrounding area, a logical point to start laying track was between Millington and Newark, thereby connecting the two towns, according to the wishes of the towns' residents.

The Monroe & Connors Construction Company of Chicago was contracted by Durant to begin work, and the company's track-laying crew (sometimes called gandy dancers) arrived, pitching its tents outside of Millington. Work was about to begin.

There had been obstacles to overcome. For example, a highway named Johnson Road, which ran in a fairly straight line, already connected the two towns, but there would be problems if Durant tried to build the Midland alongside it, as houses and private property lined the street. Outside of Millington was the Millington-Newark Cemetery, and no railroad would be permitted to build through a cemetery, so a different route would have to be surveyed. Since local farmers desperately wanted a railroad to facilitate shipping crops and livestock to market, they leased portions of their farmland to the Midland, although that brought another problem—the new railroad would pass through fields and pastures where livestock roamed, so farm gates would have to be installed for engineers to open and close to allow trains through while keeping livestock inside. There was also the problem of Clear Creek, a charming brook that wandered through the picturesque locale but presented an obstacle that would have to be crossed several times, requiring the construction of bridges.

Before a roadbed could be laid for the new railroad, preliminary grading and filling had to be done. Trains need a fairly level surface to run on, and what should have been done was a smoothing of the slight downhill slope that existed from Newark to Millington—a natural feature of land that is still called the Fox River Valley. This work, unfortunately, was skimped on (probably to save money), resulting in the hill near Newark becoming a challenge for locomotives operating along the line.

Producing a roadbed was the next step, with the railroad standard consisting of crushed stone laid to a depth of six inches. In addition to acting as support for wooden railroad ties set into it, the crushed stone, called ballast, provided drainage for the tracks. Substandard ballast was used when building the Midland, again in an effort to cut costs. Clear Creek was spanned by five small bridges made of wood; this was cheaper than using iron or steel, but it was a choice that would later lead to regret. In the fall of 1913, the spiking down of the tracks for the Illinois Midland Railway began. Compared to other railroads built 100-plus years ago, the Midland was not markedly difficult and did not involve mountains that had to be tunneled through nor deep rivers that needed massive steel bridges built over them.

Unfortunately, the materials used in constructing the new railroad were not new at all. Rails for the Illinois Midland were, in fact, secondhand lightweight ones that had been torn up from a variety of places. Even the ties the gandy dancers used to spike the rails had been used on other rail lines. At any rate, construction was completed using materials that were cheaper than what was being used on major railroads across the United States at the time.

The Midland's materials came from a variety of places. Some had been produced at a steel mill in Joliet during the 1880s, while others came from Cambria Steel in Johnstown, Pennsylvania. But rails weighing 100 pounds per yard were the industry standard a century ago (today, it is 137 pounds or better), so the Midland, which weighed as little as 60 pounds per yard or less, were too light to carry much of anything. Lightweight locomotives and freight cars with a capacity of 20 tons or less would have to be the limit for what the Midland's rails could handle.

It did not take long to build just under two miles of track that connected the two towns, and on Valentine's Day 1914, a celebration was held to celebrate the first step in building the Midland. Then financial and legal troubles came, followed by Durant's disappearance after he won a lawsuit. But what happened to him after that has remained a bit of a mystery—until now. For the first time, the remaining years of Durant's career can be explained. Previously published accounts follow him only until he left town, headed for prison. Some accounts say he fled with the money he had pocketed from selling his valueless stock to the unsuspecting citizens of the two towns; in fact, he did return, eventually winning a protracted legal battle and taking the money he had swindled from them. According to a report published in the *Earlville True Republican* newspaper on February 5, 1919, the total was nearly $10,000 (about $150,000 in today's money) when Durant left. He never came back to Illinois. In early 1920, he turned up in Louisville, Kentucky, having worked his way up to becoming manager of a trucking company in that city. A year later, he had moved east to Virginia, becoming the manager of Tidewater Lines, a company that ran trucks and buses, as well as something called Tidewater Finance Company, which might indicate that he was still engaged in his old shady habits. From there, he headed to Buffalo, New York, where his name appeared in the newspaper but not in a way he might have liked. Durant's Cadillac was used as the getaway car in a holdup, and the crooks were caught, but when the police came calling at his door with questions, he claimed the car had been stolen and that he knew nothing about the crime.

An untimely end came for Durant on August 1, 1924, while he was on a business trip to Easton, Maryland—he died of a heart attack at the age of 42. His estate, which was valued at a modest $4,000, was left to his two nieces living in Indiana. In all fairness, a couple of his ideas were farsighted for the times: an all-electric railroad from coast to coast and a beltline around Chicago connecting Rockford to Kankakee, neither of which have materialized.

North and South Railroad for Newark

Pay Nothing till Train Reaches Newark

(73)

In Consideration of the many benefits to be derived therefrom, I hereby agree to dona

Two Hundred and Fifty ——— $\frac{no}{100}$ Dollar

($250.00) as a bonus to the ILLINOIS MIDLAND RAILWAY (that is to a railway to be built by S. Durant and associates), in aid of the building of a standard gauge steam railway NORTH AND SOUTH THROUG NEWARK, Illinois, from an ILLINOIS CENTRAL connection on the south at Clarke City, in Kankakee County, to ILLINOIS CENTRAL RAILROAD connection on the north, also at Rockford, in Winnebago County, and I hereby agr to pay the same on call as made by the railway when the ties and rails are laid to Newark and the first train reach Newark from a connection with one or more standard gauge steam Trunk Line railroads south or north of Newark.

The railway to be suitable for steam power for freight and motor power for passenger service, and to properly equipped and operated and maintained immediately upon completion.

It is distinctly understood, as above stated, that nothing is payable on this subscription until the ties and rai are laid to Newark and the first train reaches Newark over said Railway from a connection on the south or nor of Newark with one or more standard gauge steam Trunk Line railroads.

This paper to be put in trust in Farmers State Bank, Newark. NOV 22 1915

Construction to commence not later than 90 days after satisfactory subscriptions are secured around Newar Ill., (weather permitting) and the railway to be operating trains into Newark within 150 days after that date at lates If construction does not start within 90 days as above, this subscription to become void and to be immediately mail back to me by the bank at that time.

To show good faith, I hereby agree that upon the railway notifying me by mail of its intention to commen construction that I will immediately deposit the above amount in cash in the above mentioned bank, as trustee, to ready for prompt payment when the first train reaches Newark, and to deposit said sum within ten days at latest aft receiving such notice.

Date 5/22/13 Signature N. J. Mor

Witness [illegible] Miller P. O. Address Newark, Ill

A piece of evidence that shows how Samuel Durant got away with his scheme should begin with the old adage, "Read the fine print before signing anything." This is what local investors, called bondholders, did not do when they signed up to help finance Durant's proposed Illinois Midland Railway in 1913. They had unwittingly agreed to pay him a substantial $10,000 bonus if he completed the railroad as far as Newark, which is less than two miles from its starting point in Millington. Durant had sold them his proposed railroad as one that would connect the cities of Rockford (to the northwest) to Kankakee (to the southeast), yet when investors claimed he had not finished the job, Durant and his lawyers argued those two cities were merely points made to illustrate where the new railroad might go if it were completed. Investors lawyered up, and the case went to court. In many ways, the case being decided in Durant's favor (and leaving investors with worthless shares of stock) came down to who had the cleverest lawyers. (Fern Dell Museum.)

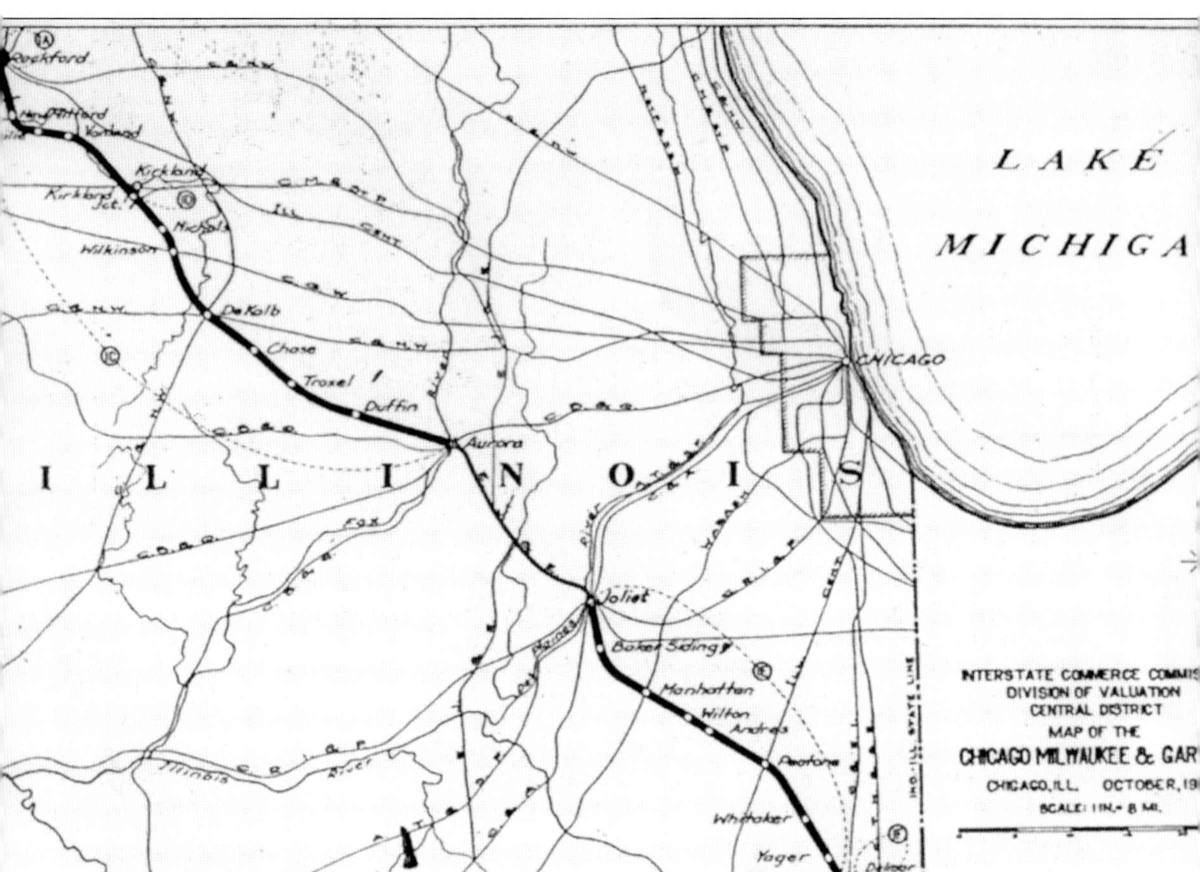

Another railroad served as the blueprint for the Illinois Midland Railway—the Chicago, Milwaukee & Gary, which was built a decade before the Midland. The Gary Line, as it became known, ran 130 miles across northern Illinois from industrial Rockford to the coal fields of Kankakee County, near Illinois's border with Indiana. The proposed route for Durant's 120-mile Midland linked the same two places. The proposed Midland would have continued south from Millington and Newark to where both routes went to the surface mines and their cheaper coal. The problem was that the Gary Line was not living up to its financial potential, so why would anyone build another railroad like it? (Wayne DeMunn.)

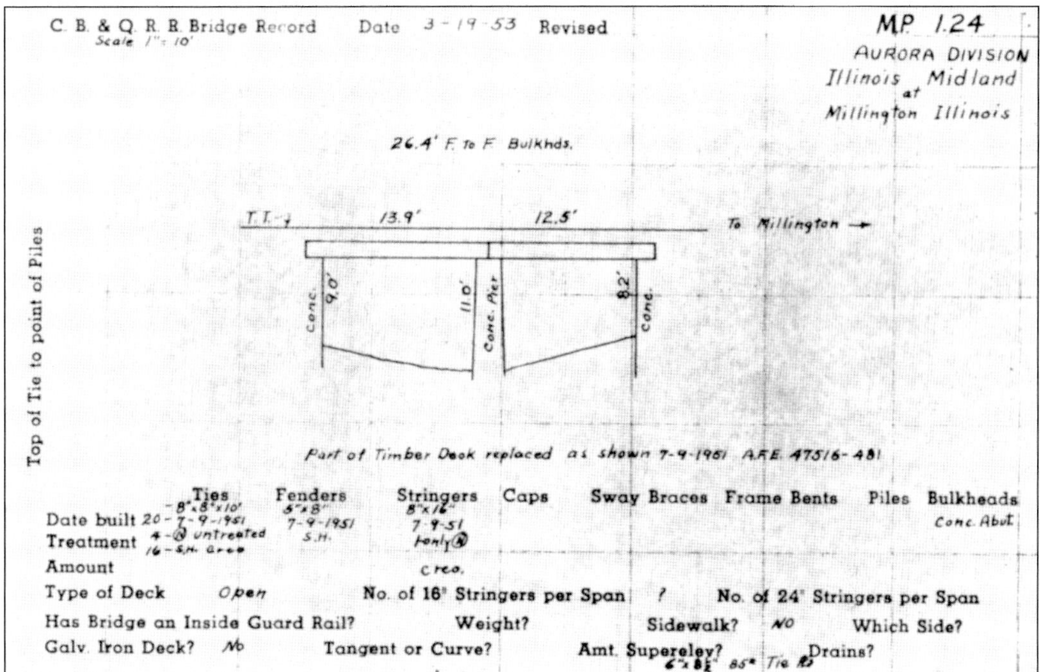

C. B. & Q. R. R. Bridge Record   Date 3-19-53   Revised

Scale 1" = 10'

MP 1.24
AURORA DIVISION
Illinois Midland
at
Millington Illinois

26.4' F to F Bulkhds.

T.T. 13.9' 12.5' To Millington →

9.0' Conc. 11.0' Conc. Pier 8.2' Conc.

Top of Tie to point of Piles

Part of Timber Deck replaced as shown 7-9-1951 A.F.E. 47516-481

	Ties	Fenders	Stringers	Caps	Sway Braces	Frame Bents	Piles	Bulkheads
	8"x8"x10'	5"x8"	8"x16"					
Date built	20 - 7-9-1951	7-9-1951	7-9-51					Conc. Abut.
Treatment	4-Ⓡ untreated 16-S.H. creo	S.H.	1-only Ⓡ					
Amount			creo.					

Type of Deck Open   No. of 16" Stringers per Span ?   No. of 24" Stringers per Span

Has Bridge an Inside Guard Rail?   Weight?   Sidewalk? NO   Which Side?

Galv. Iron Deck? No   Tangent or Curve?   Amt. Superelev? 6"x8½" 85* Tie R   Drains?

Building bridges is an important step in constructing a railroad, and it is a good idea to make them sturdy so they will stand the test of time. That did not happen with the Midland, for which cheap wood was used in places where steel girders were called for, yet the July 1951 rehab of Midland's five bridges (the railroad's longest bridge is shown here) kept the railroad going for another 16 years. (Burlington Route Historical Society.)

The unfulfilled promise to build the Midland northwest to Rockford prompted the realization that sturdy bridges, like these Milwaukee Road examples over the Rock River below Rockford, would need to be erected not only there but at Millington to permit the Midland to cross the Fox River. Although they were expensive, the bridges on the left (built in 1905) and the right (built in 1901) were still in use at the time of this 1963 photograph. (Author's collection.)

"Now *that* is a bridge," people may have said in 1910 when this impressive two-level span over the Wisconsin River was photographed at a little town called Kilbourne (it would later be renamed the Wisconsin Dells). Trains ran over the top level, while the lower level was for vehicles and pedestrians. A lack of money was primarily why the Midland never built anything like this over the Fox River. (Author's collection.)

In the beginning, the Midland had considered offering passenger service to patrons, but that idea dried up when the railroad ran out of money and ended up traversing only two miles—who would want to buy such a ticket? A hint of what might have been is shown in this photograph of the Chicago & Northwestern's Corn King Limited from September 21, 1945, with the bridge in the background belonging to the Gary Line, the inspiration for the part of the Midland's proposed route that was never built. (Author's collection.)

Water is the lifeblood of a steam engine. Most railroads built wooden water tanks along their lines a century ago to service their locomotives, but not the Midland. A standpipe coupled to a steam-powered syphon pump pulled water from Clear Creek to fill the engine's tank. As primitive as this system may seem, it worked and lasted until the end of steam on the Midland in 1959. Always operated on a shoestring budget, the little railroad excelled in finding a cheaper way of doing things; because it was surrounded by major railroads in northern Illinois, it was a constant battle. Note that the bridge where the locomotive is stopped for its water refill is one of the five built with help from the Burlington. (Fern Dell Museum.)

The conversion of an automobile to a vehicle that could run on rails as well as the highway was rather common on railroads around the country. Track inspection was a regular important duty, and this photograph from March 19, 1943, shows a comfortable, fairly new car like those that were popular at the time—except on the Midland! The Midland's vehicle was a Ford Model T (already an antique at the time this picture was taken), the subject of many photographs from the ensuing years that are featured in the next chapter. (Author's collection.)

A snowplow was a useful piece of equipment during winters on the windswept prairies of northern Illinois, but the Midland did not have one. This picture of one that belonged to the Gary Line was taken on the outskirts of DeKalb in 1954 along a stretch of open land where deep drifts of snow presented quite a challenge for trains and crews. (Author's collection.)

Around 1920, a group of curious boys watches the arrival of secondhand locomotive No. 10, soon to be renumbered Illinois Midland No. 2, which began the use of lightweight steam engines that lasted until the end of the 1950s. The unusual 0-4-0 T wheel arrangement was created by Vulcan Locomotive Works (1832–1962) of Wilkes-Barre, Pennsylvania and was known as a saddle tank type. This was a perfect fit for the Midland or any other short line railroad that moved a few cars just a short distance on lightweight track. (Fern Dell Museum.)

41½-Ton Four-Wheel Saddle Tank Steam Locomotive

GENERAL SPECIFICATIONS

Builder	Vulcan	Cab, material	Steel-Wood Lined
Gauge	4-ft. 8½-in.	Couplers	Automatic
Cylinders	14-in. x 22-in.	Boiler, type (A. S. M. E.) Code	Straight
Wheel Base	6-ft. 3-in.	Boiler, working pres. (lbs. per sq. in.)	190
Valve Motion	Stephenson	Boiler, outside diam. (smoke box end)	47-in.
Weight in working order, pounds	83,000	Firebox, length inside	46-in.
Driving wheels, diameter	38-in.	Firebox, width inside	51-in.
Tires, thickness	3-in.	Tubes, number	112
Journals, driving, diameter	7¼-in.	Tubes, diameter (o. d.)	2-in.
Journals, driving, length	8-in	Tubes, seamless, thickness	No. 11 Bwg.
Height extreme	11-ft. 10-in.	Tubes, length	125¼-in.
Width extreme	8-ft. 10-in.	Heating surface, tubes	612 sq. ft.
Length, (over bumpers)	23-ft. 6-in.	Heating surface, firebox	56 sq. ft.
Water capacity (U. S. Gallons)	1,250	Heating surface, total	668 sq. ft.
Fuel capacity, carried at rear of cab	2,000 lbs.	Brake	Westinghouse Automatic and Straight Air Brake
Rod packing	U. S. Metallic	Fuel	Coal

The Midland had found the perfect locomotive for its needs (provided its rail line never got any longer) with the saddle tank locomotive from Vulcan Locomotive Works. Its top speed was 10 miles per hour, but anything faster than 3 or 4 miles per hour could result in a derailment on the 1.962-mile route. Shown here are the specifications for such an engine—something steam aficionados would appreciate. (Author's collection.)

Maybe wooden bridges were not always the preferred choice for a railroad, but wood still saw plenty of use. A close-up view of a wooden bridge near DeKalb in 1942 shows thick and heavy wooden trestle construction, but the deck is steel and therefore fireproof. Locomotive No. 929 is a 2-6-2 prairie type well suited for branch line work. (Author's collection.)

The Midland was not the only road to have a one-stall enginehouse, although its was quaint and homey compared to this example on the Baltimore & Ohio Chicago Terminal Railroad, pictured on a snowy day in February 1963 at Faithorn Yard near the Illinois–Indiana border. If a locomotive needed repairs, working on it inside beat working outside in the winter. (Author's collection.)

Three

Workin' on the Railroad

The Midland ran its standard-gauge railroad on lightweight track using pint-sized locomotives, but what was it like being its only employee?

Despite the opening celebration held in the snowy winter of 1914, the Illinois Midland was far from a completed railroad, as it ran a distance of less than two miles between Newark and Millington. At Millington, it connected to the Burlington Railroad but did not cross the Fox River, which would have meant the necessity of building a sturdy and expensive bridge. No tracks pointed northwest to Rockford, the proposed destination of the Midland. From its opposite end, which was supposed to be built southeast to Kankakee from Newark, no railyard terminal yet existed. So, in May 1915, work commenced on one.

Materials arrived in Newark and included specially made portions of rail for switches that would enable cars to be moved from one set of tracks to another. A short portion of the trackage would run past the newly built grain elevator so cars could be loaded, and tracks would also run to a nearby lumberyard as well as a coal and oil dealership, both of which would be bought out later by the grain elevator company. An extra siding was added that led to a shed (it could not be called a roundhouse) where the Midland's locomotive would tie up for the night.

In its earliest days, the Midland's employees consisted of general manager J.P. Tutt, engineer F.A. Schaeffer, and fireman M.H. Dierzen, along with several other railroaders who used a borrowed locomotive they had labeled No. 1. The shaky finances of the Midland became apparent when, in April 1914, those few employees had to sue for back wages of $843.41. The jobs of general manager and such disappeared around the time that locomotive No. 1 was sent back to its owner in Chicago, and equipment being begged, borrowed, or leased forced the Midland to purchase an engine of its own.

Around 1920, the arrival of a used steam locomotive the Midland labeled as No. 2, a 0-4-0 T type that could be operated solely by its engineer, meant that only one man was needed for the job of running the entire railroad. That unusual 0-4-0 T arrangement was for a four-wheel saddle tank engine weighing just 40 tons and used in yard service; this was popular for situations that called for a small locomotive to shuffle cars around a railyard or an industrial location that seldom ventured onto the main line. The zeros in its designated wheel arrangement meant it had no pilot, or front-end, trucks (wheels), nor did it have rear-end trucks. It essentially consisted of a steam-powered engine mounted on a sturdy frame under which were two steam-driven axles riding on a total of four wheels. The letter "T" signified that no tender was pulled behind this locomotive, so its fuel (coal) and water were both carried on the engine. It was a very economical loco for any small railway to run, and it filled the needs of the Midland.

Vulcan Locomotive Works of Wilkes-Barre, Pennsylvania, may or may not have invented the 0-4-0 T saddle tanker, but it became their specialty and was produced in various models that served for years on railroads across the country. The one the Midland used had 37-inch drivers and could carry 2,000 pounds of coal along with 1,250 gallons of water (to be turned into steam). One distinction a Vulcan had was that while carrying the weight of its fuel and water upon the driving wheels rather than in a trailing tender, its traction was improved.

The Midland liked its Vulcans, and for most of the next 40 years, that was what it used to run the little railroad. Locomotive No. 2, although it had been purchased used, lasted well into the late 1930s before it was replaced with No. 3, but exactly what No. 3 looked like remains a bit of a mystery. During the Great Depression, it seems nobody thought to expend money taking photographs of a soot-covered little engine running on an obscure line less than two miles long. Consequently, no photographs of No. 3 are known to exist. In July 1940, No. 4 arrived. Bought secondhand like its predecessors, it was also a Vulcan. It lasted until 1959, long after other railroads had scrapped their steam engines and bought diesels. No. 4 marked the end of steam on the Midland. Efforts to save it and put it on display in a park at Newark failed. Diesel power was used through the Midland's remaining years.

For most of its life, the Illinois Midland was a one-man operation, with one worker doing every job on the railroad. The prized job, according to many railfans, was that of engineer, which took a high degree of skill in the steam era. When diesels supplanted steam, an engineer's job required a different set of skills. Only on the Midland, however, was the engineer also the brakeman, switchman, fireman, conductor, and dispatcher, as well as the track maintenance crew and master mechanic—repairs would periodically need to be made, so it helped if the engineer was also mechanically inclined.

Dick West, the sole surviving engineer of the Midland, says he liked his job, and his words were echoed by all the Midland men who went before him!

Being the engineer on the Midland involved much more than just running a train two miles back and forth between Newark and Millington. The engineer was also the fireman, brakeman, switchman, conductor, master mechanic, and any other railroad job that was needed. He was also expected to maintain the Midland's tracks, which was a job in itself, since the track was rickety and his equipment consisted of a Ford Model T converted to run on rails, which was an antique even in 1943, when this photograph was taken. Here, engineer Bill Thorsen has found himself a helper (a rare thing) to replace some ties. (Fern Dell Museum.)

The arrival of a Vulcan Locomotive Works saddle tank locomotive, which was purchased used around 1920, moved the Midland beyond having to borrow equipment in order to operate. This early view of the loco at work reveals that the new owner's name had been painted on its side, but the original No. 10 had not yet been changed to No. 2, which would have been the correct sequence for the newly built Illinois Midland Railway. (Fern Dell Museum.)

The train shed used by the Midland to store its locomotive was a homey, nondescript building. Repairs were made inside by the engineer, who was supposed to be mechanically inclined. Every morning, except on Sundays, he fired up the engine's boiler, then it would take several hours to build up a head of steam, so the engineer might use that time to do a quick track inspection. (Kendall County Historical Society.)

This is a shaky picture, but its unknown photographer risked life and limb climbing to the top of the new grain elevator in Newark on a snowy day in 1917 to capture this view. The Midland's track is shown curving across Johnson Road as it heads north toward Millington through snow-covered fields and pastures. Also visible on the left is the train shed where the railway's locomotive tied up for the night. The peak at the rear of the train shed points directly at a house located just a few yards beyond the shed, and what the photographer unintentionally included in his risky photograph was what came to be called the "spook house" in local lore. Its story is told in the last chapter of this book. (Fern Dell Museum.)

This June 1936 photograph features Midland's No. 2 locomotive doing what it did best—taking a couple of boxcars loaded with grain or corn from Newark to Millington, where the Burlington would take them the rest of the way to market (usually in Chicago). At this point, the No. 2, which the Midland had bought used, would have put in some good years for the little railroad. Of interest are the wooden boxcars that were still in widespread use at the time. Both cars were owned by the Burlington, with the initials CB&Q (Chicago, Burlington & Quincy—the railroad's formal name) painted on the sides. Boxcars made of steel, which was more durable and fire resistant, gradually replaced the worn-out wooden ones, but it was not until the 1960s that the woodies were entirely gone. (Kendall County Historical Society.)

Once a head of steam was built up in the engine, it was out of the shed and down the Midland's track—all 1.962 miles of it. The No. 4 arrived in July 1940 (no photograph of No. 3 is known to exist), and it was another 0-4-0 T saddle tank built by Vulcan Locomotive Works and bought used from a short line in Lee County, Illinois, for $3,000. The No. 4 would arguably become the railroad's most widely recognized engine as a subject for photographers, newspaper stories, and even movies. (Fern Dell Museum.)

A trip along the Midland was an adventure in itself, so in a bygone era with few regulations and a friendly engineer like Bill Thorsen, it might have been possible for a visitor to grab a cab ride, as quite a few people recall. The ever-present weeds along the right-of-way could be problematic. When crushed by the steel wheels of a passing locomotive, the rails would become slippery. That, in turn, made it hard to stop the train, especially when it was going downhill. (Millington, Illinois, Historical Society Museum.)

Shoveling coal aboard the Midland's locomotive was an arduous task, as this interesting 1940s scene shows. While most railroads had coaling towers to fill their engine tenders, the Midland expected its engineers to get out and load up the locomotive for the next day's work using just a shovel! A Vulcan Locomotive Works saddle tanker like the one the Midland owned had a capacity of 2,000 pounds of coal—that required a lot of shoveling. (Kendall County Historical Society.)

Of equal importance to loading up the loco with coal was putting in water. Here, a view of that activity shows the standpipe along the right side as well as the silhouette of a man, probably engineer Bill Thorsen, standing along the side in the lower right corner of the photograph. It is winter, so that meant the engineer would have had to chop through the ice in the creek to get at the water. (Kendall County Historical Society.)

Engineer Bill Thorsen was always friendly and willing to talk to visitors about his railroad. He is shown around 1943 and was already recognized as running a one-man operation. For example, in this image, he is about to do the job of brakeman as he uncouples a car from the engine—this particular one is a coal hopper car. (Fern Dell Museum.)

One duty of the engineer on any railroad was to "oil around" by applying oil to all the moving parts on a locomotive. A steam locomotive had quite a few moving parts, as opposed to a diesel, which was "gas 'n' go" for the most part. Larger lines employed oilers whose sole job was to make sure everything was lubricated. (Fern Dell Museum.)

A close-up of No. 4 taking on water brings to mind a story about an incident that occurred in the 1940s on the Midland during such a routine activity. Engineer Bill Thorsen was unavailable, so a substitute engineer had to be found to take his place at the throttle that day. Such a fellow was found and seemed to know the job, so that morning, he got No. 4 fired up and went on his way. Its first stop would be Clear Creek to fill up the tank with water, but the substitute engineer could not get the syphon pump to work. He was clearly doing something wrong, but what? He tried again without success. Meanwhile, the water gauge showed its supply was getting dangerously low, which meant the possibility of a deadly boiler explosion (rare, but a real possibility), so the substitute engineer dumped out the red-hot coals from the firebox. They landed on the wooden bridge deck, setting it on fire! The damage was bad, but not so bad that the bridge could not be repaired. (Fern Dell Museum.)

In an interesting side view, a comparison of size differences between a standard 40-foot boxcar and the Midland's pint-sized No. 4 illustrates why this type of locomotive could only handle a few cars at a time. Economical to operate and often needing just a one-man crew, saddle tankers like this were used on short line railroads or in industrial settings where they seldom ventured out onto the main line, since they could not pull a regular train. (Fern Dell Museum.)

The front end of No. 4 reveals a detail found on many older locomotives and freight cars and one that has been illegal for years—poling pockets. The cup-shaped indentations located on the lower corners of the loco's front end are there for a reason. When a long, sturdy wooden pole was placed in the pockets between the engine and a similarly equipped freight car, short moves could be easily made in a freight yard without having to couple and uncouple, but the practice was dangerous. If the pole splintered, it could cause great bodily harm to a switchman. (Fern Dell Museum.)

This unusual view of the rear of No. 4 was taken in the 1940s at Millington, where the Midland and Burlington connected. This was where cars were interchanged, but what is interesting is the Burlington marker post indicating where the Midland's track ended and the Burlington's began. Technically, one railroad had to get permission whenever it was using another railroad's tracks. (Fern Dell Museum.)

Millington was the junction where the exchange of cars took place between the Midland and Burlington. Loaded cars full of produce were swapped out for empties to be returned to Newark for reloading. With No. 4 on the point, a train to Newark would seldom have more than four empties. In the background are some passenger cars on a house track that were probably being stored and were no longer in use. (Fern Dell Museum.)

This particular train had a four-car limit, it seems, and since empty cars weighed much less than loaded ones, the Midland's pint-sized engine could usually handle them. The trip back to Newark was slightly uphill, so when it was not possible for the train to make it all the way up, it would have to "double the hill," which involved splitting the train in two parts and taking up each part separately. (Fern Dell Museum.)

Sometimes, an empty hopper car would be returned to the Burlington after being unloaded at Newark. The grain elevator company also owned the coal and oil business in town, so this winter scene indicates that more coal is going to be needed to heat the homes and businesses in town. Steam heat was still in use well into the 1950s. The hopper has a Burlington slogan—"Everywhere West"—painted on its side, which was free advertising. (Fern Dell Museum.)

Loading cars at the Newark Farmers Grain Elevator Company consisted of getting large quantities of grain or corn into boxcars, which was the common way of shipping farm produce a century ago. The lumber nailed across the doorway was a barrier to keep the produce inside from spilling out until the cargo could be unloaded at a flour mill or sold as cattle feed. Boxcars were multipurpose cars that could even be used to transport new automobiles. (Fern Dell Museum.)

Shoving hard now, No. 4 moves a fully loaded boxcar into position for the next day's trip down to the Burlington at Millington. The typical 40-foot boxcar could hold 20 tons of cargo, and farm produce was its most common freight. Photographs of the Midland show boxcars being filled often at the grain elevator; there were also coal hoppers and oil cars, but nothing bigger or heavier was spotted on the little railroad. (Fern Dell Museum.)

By coincidence, this image of loading cars next to Newark's grain elevator includes boxcars from both of the railroads that played a major role in the Midland's history—the Milwaukee Road and the Burlington. Boxcar No. 131128 from the Burlington is a classic wooden car from the 1920s, while Milwaukee's rib-sided steel car No. 21764 was one of many thousands created in the Milwaukee Road shops and widely used for hauling grain starting when the welded steel cars were first built in the late 1930s. It is interesting to see how railroads used the sides of their freight cars, especially boxcars that served as rolling billboards. The Burlington used the slogan "Everywhere West" to tell people that it went from the Midwest to Colorado and beyond, and the Milwaukee Road's "Route of the Hiawathas" touted the fast passenger trains it ran out to the West Coast. (Fern Dell Museum.)

With tools stuffed in the trunk and a trailer car tethered behind it, the Midland's Model T is off to work in this 1943 photograph. Although this track maintenance operation worked, it nevertheless was sure to turn heads. This was one of the eccentricities of the Midland and helped develop its growing fanbase, since it was not long before local newspapers picked up on the interesting story of the old car and the railroad that used it. (Fern Dell Museum.)

The track maintenance department on the Midland consisted of one man during most of its 53 years of existence. This photograph shows what conditions were like—the fields and pastures crossed by the railroad had gates that had to be opened and shut to let the train through and keep livestock inside. Ties had to be regularly replaced, and rails had to be repaired or replaced when they wore out. The Model T was equipped with steel wheels so it could scoot down the tracks, and it was said to run faster going backward than forward! (Fern Dell Museum.)

A look inside the office at the Newark Farmers Grain Elevator Company shows Bill Tuttle (standing) and Roy Halverson (seated), the men who ran the business. Harvey Norem (not pictured) also served as president of the railroad for 27 years. The Midland had floundered following a rocky financial start after Samuel Durant won most of its money in a lawsuit and subsequently left town. Plans to build the Midland's rails as far as Rockford and Kankakee never materialized, and in 1922, the grain elevator company took the railroad under its wing, buying it outright 20 years later. Along the way, the company bought a nearby lumberyard as well as a coal and oil business, with the railroad continuing as a one-man operation on a shoestring budget and never being extended beyond the route that was established in 1914. (Fern Dell Museum.)

Although the Midland had just one administrative executive, what did major railroad management teams look like? This picture shows Milwaukee Road employees inspecting the railroad's new locomotive as it rolled out of Baldwin Locomotive Works in Philadelphia, Pennsylvania, in late 1937. The railroad had purchased 30 of these, and the new steam engines could handle both freight and high-speed passenger trains with ease. They soon became the backbone of that railroad's fleet, so more were ordered three years later. In this scene that is much different from the office in a grain elevator on the previous page, it looks like a three-piece suit topped with a fedora was the standard for well-dressed railroad executives of the day. (Author's collection.)

The Class S-2 locomotives proved so successful that the Milwaukee Road came back for more in 1940, bringing its total to 40. Of course, this was for a railroad that spanned 10,000 miles, so the Midland's route did not need much more than its secondhand Vulcan. Still, it is interesting to see what the scene was like around 1940, perhaps at the pinnacle of steam power. (Author's collection.)

A look at the firebox when a locomotive is under construction tells a story. The Midland's firebox measured 46 by 51 inches and would have been easy to fire up by hand. However, the photograph of one of those S-2 class locos tells a different story. The workman standing next to it offers a good perspective of just how large the firebox was on that powerful locomotive. The fireman on such a locomotive would have to be able to throw shovels full of coal a distance of 12 feet for mile after mile, so a mechanical stoker was installed on all S-2 models. (Author's collection.)

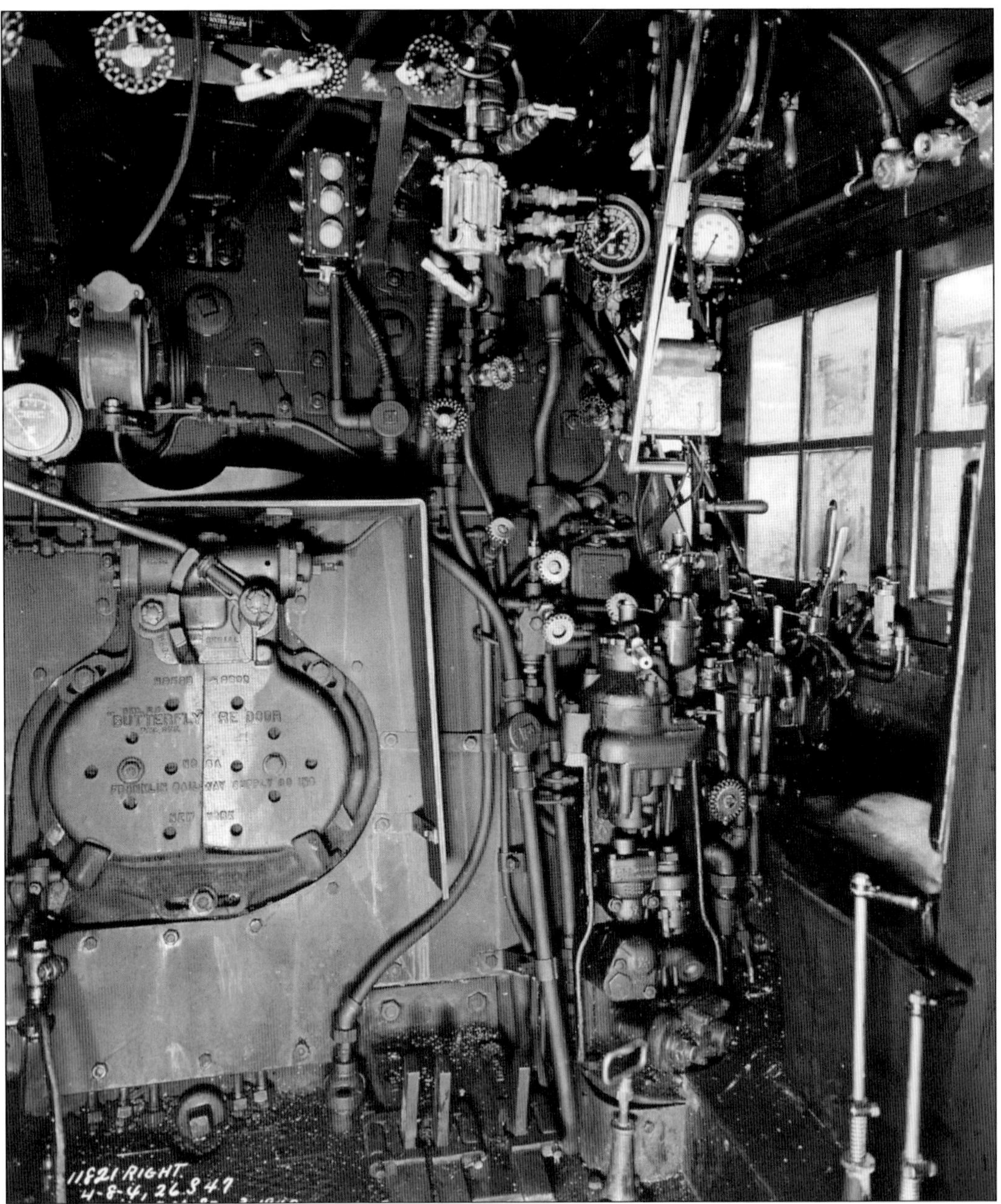

For those who have never been inside the cab of a steam locomotive or have never even seen a picture of one, this view from the engineer's side should suffice. The complicated array of valves and levers greeted an engineer as he came aboard the locomotive ready to make a run. On an S-2, the task was formidable, yet something as simple as a 40-ton saddle tanker on the Illinois Midland had the same concept—there were steam pressure gauges, a water level indicator, and the throttle (visible on the upper right). With something as big as the S-2, engineer Bill Wilkerson, who ran them, later explained in his book, *The Milwaukee S Class Locomotives*, that the advice he gave any new engineer was, "Don't make any quick, foolish moves." (Author's collection.)

A cast steel frame has arrived from General Steel Castings, incorporating the latest in foundry technology, and work is underway to turn it into a locomotive. The basis for any locomotive is a sturdy frame. Everything involved in the finished product would ride on that heavy-duty frame. (Author's collection.)

Outside the Philadelphia headquarters of the Baldwin Locomotive Works on March 18, 1940, the boiler assembly has arrived. It would take a multitude of parts from a wide range of suppliers and about two and a half months to turn it into an S-2 locomotive. Other manufacturers may have been able to do this in less time, but the results were the same, and they all started with a locomotive boiler. (Author's collection.)

The locomotive's lead truck, or pilot, is hoisted up for a better view here, revealing the location for its brake cylinders. Those small front wheels are 36 inches in diameter, which is interesting because a Vulcan saddle tanker's driving wheels were only an inch bigger. The cavernous size of the Baldwin Locomotive Works shops in Eddystone, Pennsylvania, is apparent in the background. (Author's collection.)

March 30, 1940, marked the arrival of a set of distinctive Boxpok driving wheels from the standard steel works division of Baldwin Locomotive Works. These stood an impressive 74 inches tall, which was twice the size of a Vulcan's saddle tank driving wheels. These Baldwins rode on Timken driving wheels, and when combined with multiple other parts, comprised a locomotive capable of producing 5,000 horsepower. (Author's collection.)

By early April 1940, things were beginning to come together at Baldwin Locomotive Works. The assembled parts are starting to look like a locomotive. While this particular one was being worked on, dozens of others were under construction in the same plant at the same time. Baldwin, as well as Lima and Alco, were major locomotive builders at the time, although none of them remain in business today. (Author's collection.)

The casting of the tender's floor has arrived. It, too, would have to be sturdily built, since a tender for a locomotive of this size would have to carry a lot of water and coal. To compare, the 0-4-0 T Midland engine needed 1,250 gallons of water and 2,000 pounds of coal to fill it up, while the 4-8-4 in these pictures needed a tender that carried 25 tons of coal and 20,000 gallons of water. (Author's collection.)

The front end of the locomotive is nearing completion in this picture from April 25, 1940. Steam pipes are in place to the left and right, which would send steam to the pistons. They would push side rods on the wheels, making them turn, and the engine would move. At this point, a few more baffle plates need to be added, and lagging is being placed around the boiler's perimeter. All steam locomotives were alike in many respects. (Author's collection.)

A finished front end shows a state-of-the-art (for 1940) locomotive with certain appliances all engines had in common. First is a headlight, which would be turned on day or night for better visibility. The road number of the engine, in this case No. 232, can be read from the front as well as the sides, where the numbers are illuminated. Ladders on either side make it easy for a crewman to climb up for an inspection or quick repair, and near the top on either side are marker lights meant to be illuminated when the locomotive was pulling a train. Something a little different on this new engine is an air horn, visible on the upper left corner; the locomotive is also equipped with a traditional steam whistle, so an engineer had the option of using either. The brass bell on the right was also used to warn people of the approaching locomotive. (Author's collection.)

Another innovation was roller-bearing equipped axles, one of which is shown here. Railroad locomotives and rolling stock were being built with roller bearings by the 1940s, although older friction-bearing axles continued to be used for decades. The friction bearings needed regular greasing applied to journals or they could overheat, which could lead to a "hot box" that would result in a derailment. (Author's collection.)

An unusual rear view of a nearly completed 4-8-4 permits a peek inside its vestibule cab, where the fire doors are visible along with many gauges and valves. Below the deck, the circular fitting is the stoker mechanism; beneath that sits the buffer plate, and protruding just above the shop floor is the drawbar, which is how the tender would be connected to the locomotive when they were finally put together. (Author's collection.)

The finished product has emerged from Baldwin Locomotive Works outside of Philadelphia, Pennsylvania, in 1940. Views of both sides are shown because they vary slightly from the engineer's side (above) to the fireman's side (below) and are called builder's photographs because people and background buildings have been cropped out. These measured 116 feet long, while a locomotive like the Midland's Vulcan 0-4-0 T was slightly over 23 feet in length. The Baldwin had 5,000 horsepower, while Vulcan did not advertise the horsepower of its engines. (Author's collection.)

The Milwaukee Road owned 40 of these S-2 class engines and was apparently pleased with their performance across its system, where it was said that an S-2 could easily pull 18 passenger cars, although they were mostly used in high-speed freight service. The S-2s quickly became an endangered species when diesels took over, and by 1955, they had been scrapped. Not a single one was saved on the Milwaukee Road. Meanwhile, on the Midland, steam-powered railroading soldiered on! (Author's collection.)

Diesel engines began to take over on railroads everywhere by the 1950s, around the time this staged publicity photograph was taken. Railroading had changed, but the crew for a typical freight train was the same as it had been for decades. Up front were the engineer, front-end brakeman, and fireman, and at the rear were the conductor and one (or possibly two) rear brakemen. The total crew was usually five or maybe six people, depending on the locale (mountainous or not); on the Midland there was only one crew member. Of course, a "real" railroad covered maybe 100 miles between division points, which was enough for crews to put in an eight-hour day according to union rules. A run of under two miles did not take anywhere near that time. Here, the conductor is climbing down from his caboose and being greeted by a railroad official, while his brakeman is taking down the markers—which were metal paddles during the day but kerosene lamps by night—indicating that this run is now over. (Author's collection.)

Inside the depot at Millington was the station agent's cozy, cluttered office. The job of station agent, even at a remote location like Millington, with its connection to the Midland, was an important part of working on the railroad. In addition to selling passenger tickets for the trains that called there, the agent was responsible for filling out paperwork, handling a small amount of freight and mail, and passing train orders to train crews via telegraph. As the agent for a major railroad like the Burlington, he took on the role of big brother, writing up waybills to accompany the few loaded freight cars contributed by the Midland each day. Passenger service ended on the Burlington's Fox River Line in 1952, and the agent's job was clearly in jeopardy. Modern signaling and radio crew communication brought an end to a colorful era, and depots along the line were demolished in the 1960s. (Burlington Route Historical Society.)

There is nothing like a passenger train roaring by at speed to make a tower operator realize he had better have his switches correctly aligned for its passing. This job required a man to be alert at all times, as most towers were manned around the clock, and trains could pass by at any time of the day or night. Although no such requirement existed at Millington, an agent there might be called to fill in for someone at another location, so he had better know the job. (Author's collection.)

On duty over 40 years ago at Chicagoland's Tower B-17, operator Bud Brumfield kept the trains rolling by having control of the CTC (Centralized Traffic Control) board. By then, telephone and radio had replaced telegraph communication, but it was kept as a backup, and an operator was expected to know it. As Brumfield said of the operator's Morse code test, "Either you knew it or you didn't." Railroaders like the station agent and tower operator were living proof not everyone workin' on the railroad got to sit up front and run a train. (Author's collection.)

Four

Little Railroad, Big Fan Club

How could a little railroad that struggled so mightily could become newsworthy, and what made the Midland famous compared to other roads?

The Midland being such a short railroad (only 1.962 miles long) attracted fans with its little locomotive running a train of just a couple of cars down rickety rails through farm fields and a cow pasture. It was a David vs. Goliath story told in railroad language, and newspapers picked up on it beginning in the 1940s and continued to cover it into the 1960s and until the railroad's end. It was said that if the Midland delivered just one car per day, it still made money. That is understandable when the railroad had only one employee! It ran successfully because of the extraordinary efforts of its engineers, who were friendly, hardworking men willing to talk to people (including newspaper reporters) about their railroad. Stories about the Midland soon spread far and wide. A film crew even descended upon the quaint line and shot some black-and-white newsreel footage of the little locomotive in action. Surprisingly, a couple of minutes of this film still exist; unfortunately, it is a silent film, but with a real soundtrack, it would have been possible to hear the whistle on old No. 4 as well as its clanging bell.

The Midland's fame grew when a big-city newspaper, the *Chicago Times*, got wind of the railroad 50 miles southwest of the city and sent out a reporter, photographer, and sketch artist to get a story that ran as a Sunday edition feature on November 7, 1946. The human-interest story was titled "One Man Railroad" and featured Bill Thorsen, the Midland's engineer and jack-of-all-trades. It focused on Thorsen running the railroad and included some of Thorsen's life story rather than the oft-told tale of the railroad and its early financial turmoil (Samuel Durant was not even mentioned). Instead, it told how farmers and their Newark Farmers Grain Elevator Company came to own the railroad and how Thorsen ran the train.

Thorsen, in his affable way, explained to reporter Judith Allen that he had not started out as a railroader but a mechanic working in a local garage who was hired by the grain elevator company to maintain its first engine, and he had been at it ever since. A veteran of World War I, Thorsen and his wife, Susie, and daughter, Bertha Jane, lived in their Newark home just a block from where he had grown up.

Allen took readers for a trip on typical day down the Midland, which would begin with Thorsen firing up No. 4 inside its shed at Newark; while it worked up a head of steam, he would drag out the Midland's Ford Model T (also maintained by Thorsen), equipped with steel wheels, for a quick track inspection run to Millington. He might have to pause along the way to fix a fence, cut some weeds along the right-of-way, or shovel snow (in winter). By midafternoon, No. 4 was ready to make its two-mile run, pulling one or two boxcars loaded with grain out of the yard alongside the elevator. Allen told how the little train "bumps, weavers and screeches along, with Bill leaning anxiously out of her cab, guiding her gingerly down the uncertain tracks." The return trip might have an empty boxcar or two for loading at the elevator and maybe one filled with lumber or fertilizer or a hopper car filled with coal. Many people who read this story wanted to see the Midland in action, and Thorsen did not mind the attention at all.

A high point was reached when the already-famous Midland made it into the *Guinness Book of Records* as the World's Shortest Railroad.

Even before it made the Guinness book, the dubious honor of being called the shortest railroad in the state of Illinois had not been without its challenges. According to a June 24, 1953, story in the *Galesburg Register-Mail*, residents in western Illinois's Knox County had thought for years their Galesburg & Great Eastern held that distinction. That railroad linked a strip coal mine with the Burlington, but at a little over 10 miles in length, it was not even close to the shortest line!

Melvin Severson was the featured engineer in a *Chicago Tribune* article from April 17, 1956. Again, readers got a glimpse into working on the Midland. The day that reporter Thomas Morrow visited the railroad, Severson had three loaded cars to haul from Newark to Millington and three empties waiting for him on the return trip. Despite being much lighter, his little engine could only manage two boxcars at a time going uphill, so he would have to make two trips that day. Severson did not complain about the extra work. He added, "You have to watch for calves . . . when they get on the track, you have to get down and put them off."

Perhaps the last time the Midland's celebrity locomotive No. 4 made the news was in the fall of 1958, when the Chicagoland suburb of Palos Hills announced it was organizing an Indian summer steam train excursion on October 12 through the colorful foliage of the Fox River Valley. The Burlington would bring a 4-6-4 steamer out of retirement to pull a train of dome cars and the like, departing from Union Station and then heading out and around northern Illinois for a day. A special stop was scheduled at Millington for a demonstration of the Midland's No. 4 steam engine. The fare was $6.75 for the excursion, and those who attended were lucky to see No. 4 in action—a year later, it was withdrawn from service due to insurmountable mechanical problems.

Steam ended on the Midland at the end of the 1950s, and diesels were in use for the remainder of the Midland's years, but its fans remained loyal.

This is probably the most well-known photograph of the Midland in action. It features engineer Bill Thorsen and locomotive No. 4 in a setting that illustrates a typical workday. Opening and closing farm gates was part of the job, but so was being the locomotive's engineer, fireman, switchman, conductor, master mechanic, track repair crew (note the lightweight track in the foreground that seemed to be always needing repair), and more. The photographer of this 1943 scene remains unknown, but sometime later, the title "One-Man Railroad" was added, and the photograph circulated widely. Thorsen was always ready to talk to people about the Midland whenever they stopped by to watch the action. Newspaper reporters could count on him for a story, too—as a sort of goodwill ambassador for the Midland, he helped to spread its fame far and wide. (Fern Dell Museum.)

The Midland engineer Bill Thorsen posed with the railroad's Model T on one of the five bridges the Midland crossed between Newark and Millington. It is possible this is the bridge at milepost 1.24, which is shown in the blueprint below and is the largest of the five. When Thorsen and the Model T were at work, they caught the attention of locals, and in the 1940s, newspapers like the *Aurora Beacon-News* began to feature short stories about the railroad. (Fern Dell Museum.)

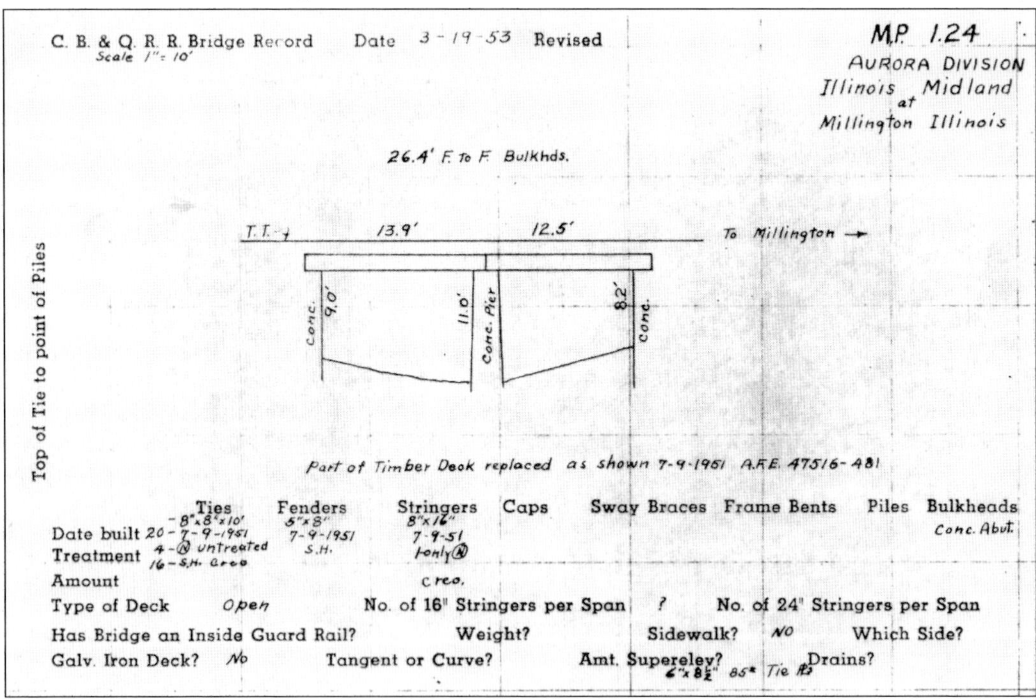

C. B. & Q. R. R. Bridge Record   Date 3-19-53   Revised

Scale 1" = 10'

M.P. 1.24
AURORA DIVISION
Illinois at Midland
Millington Illinois

26.4' F. to F. Bulkhds.

T.T. 13.9' 12.5' To Millington →

Conc. 9.0'   11.0' Conc. Pier   8.2' Conc.

Top of Tie to point of Piles

Part of Timber Deck replaced as shown 7-9-1951 A.F.E. 47516-481

	Ties	Fenders	Stringers	Caps	Sway Braces	Frame Bents	Piles	Bulkheads
	20 - 8"x8"x10'	5"x8"	8"x16"					
Date built	7-9-1951	7-9-1951	7-9-51					Conc. Abut.
Treatment	4-Ⓝ untreated 16-S.H. Creo	S.H.	1-only Ⓝ					
Amount			creo.					

Type of Deck Open   No. of 16" Stringers per Span ?   No. of 24" Stringers per Span

Has Bridge an Inside Guard Rail?   Weight?   Sidewalk? NO   Which Side?

Galv. Iron Deck? No   Tangent or Curve?   Amt. Superelev? 6"x8½" 85* Tie   Drains?

The bridge over Clear Creek at milepost 1.24, as designated by the Burlington Railroad's Aurora Division, Illinois Midland at Millington, Illinois, was 26.4 feet across and is noted to have been built in 1951 (or more likely rebuilt from an earlier version) in this drawing made on March 19, 1953. Among the blueprint specifications is one in the lower left corner that reads: "Galvanized iron deck?" and the answer, "No," which would eventually lead to the Midland's downfall in 1967. (Burlington Route Historical Society.)

The Midland was always a railroad with a fan club; even around 1920, when a "new" locomotive arrived, people came out to see it. Note the curious young boys getting a closer look and the older men who were likely discussing the prospects for the future of what had just arrived. The secondhand locomotive the railroad bought would last into the late 1930s. A Vulcan 0-4-0 T locomotive was a perfect fit for the Midland provided the railroad did not add any more track or try to make it haul any heavy trains. (Fern Dell Museum.)

AUTHORIZED SIGNATURE OF

ROY ASH.

Farmers State Bank of Newark
NEWARK, ILLINOIS

SIGNATURE *Roy Ash*

Engineer on the Midland.

ADDRESS Newark, Ill.

DATE Sept. 8, 1925
Poucher—Minneapolis Form 11 x

AUTHORIZED SIGNATURE OF

S. G. Durant

FARMERS STATE BANK OF NEWARK
NEWARK, ILLINOIS

SIGNATURE *Illinois Midland Ry*
By S. G. Durant

ADDRESS *Newark, Ills.*

DATE *Jan. 26, 1917*
KIMBALL-STORER CO., MINNEAPOLIS FORM 11 X

Original bank transactions for the earliest days of the Midland still survive today. The first of these, dated 1925, is written to Roy Ash, who would have been an engineer on the Midland during its earliest days. The second one, written and signed by Samuel Durant, dates to January 26, 1917. Pieces of Midland memorabilia like these, as well as original stock certificates, still exist today; they have no monetary value but are considered collectibles. (Fern Dell Museum.)

Boxcars of grain, such as this old wooden outside-braced Burlington car with a 20-ton capacity, were a common sight along the Midland for decades. People who lived along the line saw them every day, and this picture was taken looking east along Crimmins Road where it crossed the Midland, the perfect location for an action photograph. (Fern Dell Museum.)

Leaving the engine shed every morning, the Midland's old No. 4 would be on its way, and engineer Bill Thorsen would usually be at the throttle. With such lightweight rail, the amount of freight that could be moved over the Midland's tracks was extremely limited. Even larger 50-foot boxcars were not permitted, and a one-man operation did not call for a caboose, which would have been a typical fixture on other railroads at the time. (Fern Dell Museum.)

Climbing aboard Midland No. 4 was something engineer Bill Thorsen would do a dozen times a day. He would start his day once he had gotten a head of steam fired up in the locomotive's boiler, then head over the to grain elevator's yard to pick up cars. From there, it was down the line to Millington, but he would have to open and shut a couple of farm gates to let the train pass through pastures and fields. It was interesting railroading to watch! (Fern Dell Museum.)

The trip down to the Fox River was a scenic one, with the train rolling across one of the five bridges over Clear Creek or a tributary. The area was a bit soggy, and with a water table that was quite near the surface, vegetation abounded. The right-of-way running between the two towns had been leased from various landowners, some of whom installed gates in the fences they had built to keep livestock inside their pastures. (Fern Dell Museum.)

A good view of locomotive No. 4 taking on water reveals what a simple but effective job it was, although the remote location for this activity made photographing it a bit of a challenge. Thomas Morrow, a reporter for the *Chicago Tribune*, made the trip to see the Midland in action and wrote about it in his column on November 7, 1952. With a readership like that of the *Tribune* (the largest newspaper in Chicago), more people learned about the Midland. Its fan club continued to grow, and people who made the trip to Newark to see it in action might even get a cab ride in No. 4. On the day Morrow interviewed engineer Bill Thorsen, a fireman named Mel Severson, who would soon become the next engineer, was with Thorsen. Not one to complain about his engineering job, Thorsen did vent his frustration with the grain company's board of directors, saying, "They should not be allowed to own a locomotive without having water for it." No reply to that remark was ever made by the board. (Millington, Illinois, Historical Society Museum.)

Once in a while, a helper might assist the Midland's engineer as he was moving cars in the small yard next to the grain elevator. Boxcars, in particular, had to be in just the right location for loading grain. On a larger railroad, that was usually the job of the brakeman, and it could be a dangerous one, as a mistake could cost him his life. Hand signals were used to communicate; at night, swinging a lantern made the signals visible to an engineer. (Fern Dell Museum.)

Engineer Bill Thorsen was always willing to pose proudly with locomotive No. 4. As was the case with all the locomotives that ran on the Midland, it was bought used. Built by Vulcan Locomotive Works in 1924, it arrived at Newark in July 1940; the unusual 0-4-0 T wheel arrangement was well suited to the Midland's uneven and wobbly track. Thorsen grew accustomed to dealing with the public as well as reporters out to cover the kind of human-interest story readers would enjoy. (Burlington Route Historical Society.)

Chugging by on the Midland's tracks is its only engine, while in the background sits the Millington depot. It was there that cars were interchanged with the Burlington on its Fox River Line, so railfans might get to see some action every afternoon. Also worth noting is how overgrown with weeds the Midland's tracks became during the summer months. With one train a day, not much kept the weeds from running wild. (Millington, Illinois, Historical Society Museum.)

It is never a good idea to stand in the middle of the tracks to get a picture of an oncoming locomotive, but judging by the fact that the headlight is not illuminated (and the train is sitting on top of the Clear Creek bridge, where water was taken up), the No. 4 was probably not moving. Of course, even if it were, there would be some fans who would take that risk knowing the train never ran at more than three to four miles per hour, so they could easily get out of the way. (Fern Dell Museum.)

Speaking of speed—or the lack thereof—on the Midland, the same fans who appreciated the railroad and its quirkiness would have been impressed by what major railroads of the day were doing with it. Streamlined trains designed in the 1930s began making headlines across the country as people lined the tracks to watch them go speeding past. One of the first was the Burlington's Zephyr, which set a speed record between Denver and Chicago using a stainless-steel diesel-powered train that raced past an admiring crowd in nearby Aurora. In 1935, the Milwaukee Road introduced its steam-powered Hiawatha, which was so successful that by 1938, a stronger locomotive was needed to pull all the extra passenger cars it carried. That engine, which had a 2-6-4 wheel arrangement known as a Hudson (or Baltic, to a Milwaukee man) pulled passenger trains every day at over 100 miles per hour. (Milwaukee Road Historical Association.)

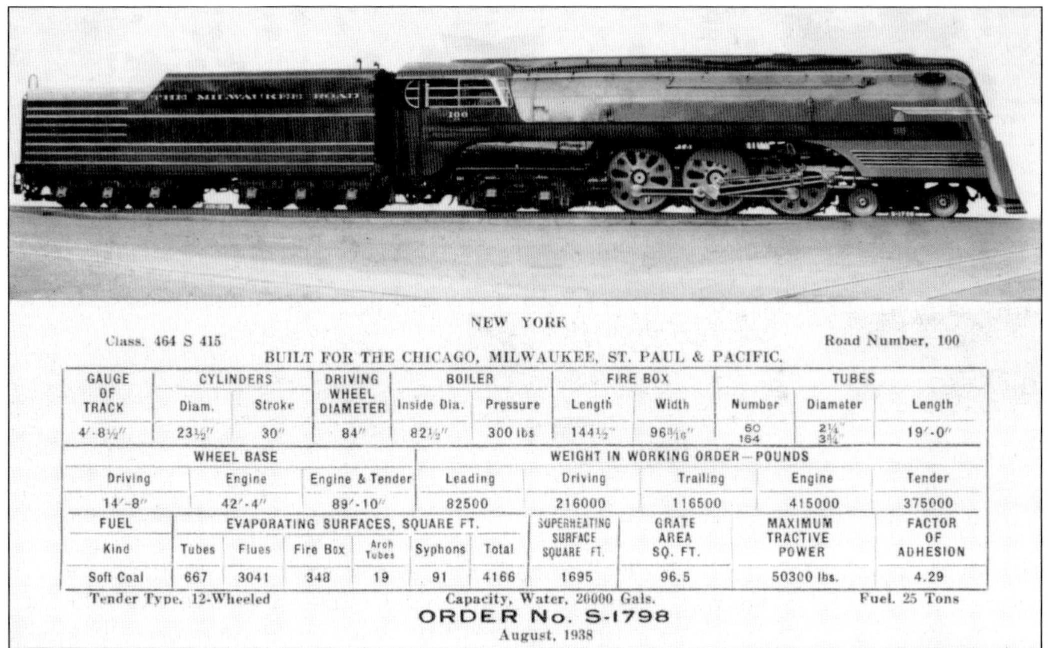

NEW YORK

Class. 464 S 415 — Road Number, 100

BUILT FOR THE CHICAGO, MILWAUKEE, ST. PAUL & PACIFIC.

GAUGE OF TRACK	CYLINDERS		DRIVING WHEEL DIAMETER	BOILER		FIRE BOX		TUBES		
	Diam.	Stroke		Inside Dia.	Pressure	Length	Width	Number	Diameter	Length
4′-8½″	23½″	30″	84″	82½″	300 lbs	144½″	96 3/16″	60 164	2¼″ 3½″	19′-0″

WHEEL BASE			WEIGHT IN WORKING ORDER—POUNDS				
Driving	Engine	Engine & Tender	Leading	Driving	Trailing	Engine	Tender
14′-8″	42′-4″	89′-10″	82500	216000	116500	415000	375000

FUEL	EVAPORATING SURFACES, SQUARE FT.						SUPERHEATING SURFACE SQUARE FT.	GRATE AREA SQ. FT.	MAXIMUM TRACTIVE POWER	FACTOR OF ADHESION
Kind	Tubes	Flues	Fire Box	Arch Tubes	Syphons	Total				
Soft Coal	667	3041	348	19	91	4166	1695	96.5	50300 lbs.	4.29

Tender Type, 12-Wheeled — Capacity, Water, 20000 Gals. — Fuel, 25 Tons

ORDER No. S-1798

August, 1938

A builder's photograph of the 1938 Hiawatha, touted by the Milwaukee Road with its slogan "Nothing faster on the rails," hit record-setting speeds of 120 miles per hour—with a steam engine! Included here are the specifications, telling the story of speed on the rails in language only steam aficionados might understand. However, the diesel would ultimately win the contest, and the stylish steam-powered Hiawatha locomotives were all scrapped by 1952. (Author's collection.)

The Midland, had it been built to the original plans from 1913, would not have had high-speed service, but it might have had something like the train in this image from May 30, 1947. The location is Monroe, Wisconsin, on a line that ran to Mineral Point, featuring a "mixed train" that carried both passenger and freight cars on a comparable 82-mile branch line. Such an arrangement might have worked on the Midland, since it could have covered both types of service rolled into one. (Author's collection.)

What type of locomotive would have been the choice of power on the Midland that was envisioned in 1913? It might have been one with a 2-8-2 wheel arrangement, which was classified as the L-2b on the Milwaukee Road. The track needed to support its weight would have needed to be heavier, but these workhorses were used everywhere—in the yards and out on the main line pulling hefty amounts of freight. (Author's collection.)

What the Midland did best was shuffle cars around a small railyard, then deliver them—after they were loaded—to another railroad less than two miles away. In that, it was successful, even if the train occasionally consisted of only one car. The simplicity of a one-man railroad somehow appealed to people's sense of humor, and it was kind of amazing how it all got done. (Kendall County Historical Association.)

Just a few miles from the Midland and its steam engines sits the town of Sandwich, where a talented railfan named Augie Otto had a machine shop and built small working models of engines based on scaled-down blueprints in his spare time. This early-1950s photograph shows a crowd of curious onlookers and one of his creations, which was probably on display at the annual Sandwich Fair. Note the lettering on the tender—"O.T.&O.," as well as the number 1951 under the cab windows—indicating the year it was built and probably when the photograph was taken. Otto eventually built a larger steam-powered model based on Burlington blueprints that took five years to complete (all the parts had to be custom made), and it was so successful that it still pulls carloads of people around a large circle of track every year at the town's fairgrounds. (Fern Dell Museum.)

The Midland never owned a caboose, but if it had, the style of caboose that might have been a good fit for the World's Shortest Railroad could have been a "bobber" type, like this one. Note the two axles (rather than the standard four), which made this little caboose the right size for short trains on a short line. It was a money-saver, too; there was little need for crew comforts, because the train would seldom venture far from home. The word "bobber" said it all, because it was so lightweight that it tended to bounce along on the end of trains and also had a tendency to derail. Essentially, it was not safe, so by the time of the World War I era, when the Midland was just starting, most railroads were already outlawing bobber cabooses, and by the 1930s, they were gone. This example was 17 feet, 9 inches long and built in 1892. (Author's collection.)

Primitive? Sure, but this is what the inside of many cabooses looked like after they had put in many miles of service. They often did not receive much maintenance, so riding in one (as well as often having to spend the night there) was a little like staying in a rustic cabin somewhere in the woods. This one's potbellied stove was used for both heating and cooking. (Milwaukee Road Historical Association.)

As opposed to the previous image, any railroad would rather have the public see a photograph like this one taken at the Milwaukee Road shops on July 29, 1939. Here, a brand-new bay window caboose has just been built and would soon enter service. In time, it could end up looking like the previous picture, but for a few years, it was a rolling motel room with some of the comforts of home. The Milwaukee Road and the Baltimore & Ohio pioneered the bay window caboose in the 1930s, then many other railroads followed with similar versions of this type of car, which was proven to be safer for crews. (Milwaukee Road Historical Association.)

The Midland never had a caboose, but they were invented, so to speak, before the Civil War, so by 1879, they looked like this. The 30-foot-long caboose shown here in a railroad staged publicity image would not have so many neatly dressed crewmen aboard; it was one of 121 built by Dubuque Shops in Iowa that featured a side door on one end for loading and unloading small amounts of freight. Something like this might have been ideal for the Midland. (Author's collection.)

The arch bar truck was found under nearly every wooden car throughout the 1800s, and it is shown on the caboose in the previous picture. These wheel sets were easy for a shop's crew to knock together, as they relied on wooden braces and bolts, which constantly needed tightening to hold them together. Note the original journal box (left) and its replacement (right), which was a newer variety. (Author's collection.)

Dieselization came to the Midland in 1959, after locomotive No. 4 just plain wore out. Repairs for the old locomotive, which was built in 1924, would have cost more than what the Midland paid for it when it was purchased in 1940. The end for No. 4 also marked the end of steam power on the Midland, as the rail's first diesel engine was bought secondhand from the Burlington. This is a builder's photograph from 1932 of the 400-horsepower locomotive fresh from the shops of the Whitcomb Locomotive Company (1896–1952), located in nearby Rochelle, Illinois. Technically, this particular model switcher's engines burned gasoline rather than diesel fuel, but the concept was the same—internal combustion engines turned onboard electric generators that sent power to traction motors on axles that turned the wheels and made it run. By the time it arrived on the Midland, this locomotive had seen better days, but age was not the whole problem; at 65 tons, it was too heavy for the lightweight track. (Burlington Route Historical Society.)

The Burlington had renumbered the Whitcomb switcher several times over its three decades of existence before it came to the Midland. Formerly No. 9120, it should have become Midland No. 5, but that did not happen. "It tore up the tracks," according to folks who remember it. By 1964, it was headed for the scrap yard, still wearing Burlington paint and the same number after both of its generators burned out. The Midland needed a replacement locomotive—and soon! (Millington, Illinois, Historical Society Museum.)

What might have worked for the Midland looked similar to the locomotive pictured here, because at 44 tons, it was in the range of what was safe to use on the lightweight track. Built by Whitcomb in October 1941, it was powered by two Caterpillar diesel engines producing 360 horsepower, which would have been more than enough for the Midland's modest needs. However, this particular model was somewhat rare, and none would have been for sale in 1964. (Sam Carlson.)

With high hopes, a 35-ton diesel arrived on the Midland in 1964 after it was purchased secondhand from the Quincy Soybean Company. It had been built by the Plymouth Locomotive Company (1914–1999) of Plymouth, Ohio, which produced 7,500 similar lightweight locomotives; an estimated 1,700 of these still are still in service today. Grain elevators were high on the company's list of customers, as well as lumber companies, clay pits, brickyards, quarries, steel mills, manufacturing plants, plantations, refineries, all types of mines, and even lettuce farms. By the time the Midland got its locomotive, the original engine had been replaced with a Caterpillar diesel, a practice that continues today. Basically, all the Midland had to do was paint over the name with some matching yellow paint. (Fern Dell Museum.)

Shown here in the cab of what should have been number Midland No. 6 is the railroad's last engineer, Dick West. For no apparent reason, the little Plymouth diesel never got a number painted on its side. Note the lettering on the locomotive's cab, with the name Richard West digitally added to the photograph in recent years to identify him. This was not stenciled on the engine, although on some railroads, a person's name might be painted on one of its locomotives as a way of honoring their years of service. West's years of service with the Newark Farmers Grain Elevator Company spanned from 1964 to 1968. When he was hired, he was the "young guy," even though he was in his 30s at the time. He soon discovered that the old-timers often stuck him with the jobs that might be considered a bit dangerous; he also said several other grain company employees were capable of running the train, but most of the time it was him. (Millington, Illinois, Historical Society Museum.)

The Midland had a way of getting the most out of its locomotives (which railfans loved), all of which were bought used, over its 53-year existence. No. 4, for example, racked up 35 years on the rails until repairing it would have cost more than it was worth. Few steam locos could compare with the X-999, a locomotive that looked like something from the Civil War that was too ugly to save. It was built around 1887 by Grant Locomotive Works (1867–1895) as No. 737, bought by the Milwaukee Road and renumbered No. 577 in 1899, then renumbered No. 527 in 1913, and renumbered again in 1936 as X-999. When it was photographed in June 1938, the X prefix meant this engine was now assigned to nonrevenue service shuffling cars around the yards. Still, it remained in service until 1947, when it was scrapped—that is an amazing 60 years! (Author's collection.)

Five

Two Tales of Two Cities

Millington and Newark were linked by the Midland, but what stories can each town tell to visitors who think they are alike?

Neither Newark nor Millington could be called a city today, nor were they anything like that in previous years. What existed a century ago was the potential for each town to become much larger than it ever did, and railroads played a role in that story.

Both towns began around the same time in the 1830s, with farming quickly becoming the livelihood of most new arrivals in and around the area. That enterprise continues today. However, Newark and Millington were always more than farming communities, so a look at their similarities and differences might help to explain their varied tales.

Education in the area goes back to the time when rural public schools often consisted of one-room schoolhouses. One teacher taught the "three Rs" for all grade levels. However, the demand for a higher standard of education existed, which helped lead to the creation of the Fowler Institute in Newark in 1855. Several "professors" taught a variety of subjects at this private school as well as others like it. Newark's Fowler Institute was a large, three-story wooden building that was well attended until it was destroyed by fire in 1880 and never rebuilt. Eventually, Newark's high school filled the void for secondary education.

Millington had no adversity to education, replacing its one-room schoolhouse with a sturdy two-story brick building named Millington School with two classrooms in 1922. That structure is still standing today, although it is no longer used as a school. Declining enrollment forced its closure in 1975, so students are now transported to one of two nearby districts. The Millington, Illinois, Historical Society has been housed in the former schoolhouse since 2017.

Also of some historic interest is the lack of saloons in Newark. Early on, there were saloons in the town, but Newark's first experiences with them had been bad—it was like the Wild West for a time. A vote was taken, and citizens decided that prohibition was the way to go; the saloons were closed down, and liquor sales were prohibited in the town to the point that drugstores were forbidden to sell alcohol "even for medicinal purposes." No such laws existed in Millington.

Stagecoach lines gave way to railroads, as previously explained, and getting rail service to promote the growth of any town became very important around a century ago. In 1871, when Millington got the new railroad Newark had been courting (the Ottawa, Oswego & Fox River Valley), hurt feelings were only part of the result.

At the same time, a proposal was made to build a narrow-gauge line from Chicago to Millington (50 miles), then continuing southwest to Muscatine, Iowa. The new railroad would be known as the Chicago, Millington & Western. That name alone shows the growing importance of Millington as a rail center in northern Illinois. Rumors spread that the little town along the Fox River could soon become "the next Chicago."

Unfortunately, as is often true with ill-conceived railways, the bottom line was money—or the lack of it, in this case. A narrow-gauge line was cheaper and easier to build, but the fledgling

Chicago, Millington & Western (CM&W) laid track only a few miles out of Chicago to a gravel pit before it ended. The rest of the proposed railroad never made it off the drawing board.

Through the last two decades of the 19th century, a few other attempts were made to build rail lines across northern Illinois with Newark featured as one of the stops along the way. One plan from 1900 called for a standard-gauge line starting in Joliet that would run to Newark by way of Plainfield; despite high hopes, nothing ever came of it.

In 1913, the arrival of Samuel Durant and his plans for the Midland to "put Newark on the map" were initially greeted with enthusiasm, but financial turmoil followed. The whole thing ended up in court, with railroad bondholders (the farmers and investors, mostly from Newark) hiring lawyers to go up against Durant and his lawyers. Trying to arouse some animosity between the two towns, Durant made public statements and even published an open letter to the citizens of Newark, saying, "Millington controls your lawyers and are working their own game at your expense." Whether there was any truth to his statement is not important, but Durant left town with a court-ordered settlement of nearly $10,000 from a lawsuit that may have come down to who had more clever lawyers, as stated earlier in this book.

The Burlington linked the two towns with the outside world. A junction at Millington was the point of interchange for the Midland's freight cars, and its depot had a Burlington agent who sold tickets to the few passengers who might be traveling by train to one of the stops along the Burlington's 57-mile route. Although freight business on the Fox River Line was good, passenger patronage continued eroding, so by the late 1920s, the railroad was looking for ways to economize the service.

Burlington's substitute for passenger trains was a single motorized coach known up and down the line as the Dinky. An oversized gas-powered interurban car, the Dinky could carry 30 passengers as well as sacks of mail and milk cans. Longtime residents can still recall riding the single-car train and hearing its high-pitched air horn tooting a warning at country grade crossings along the way. Only one serious accident marred its long history of service, and it is discussed in this chapter.

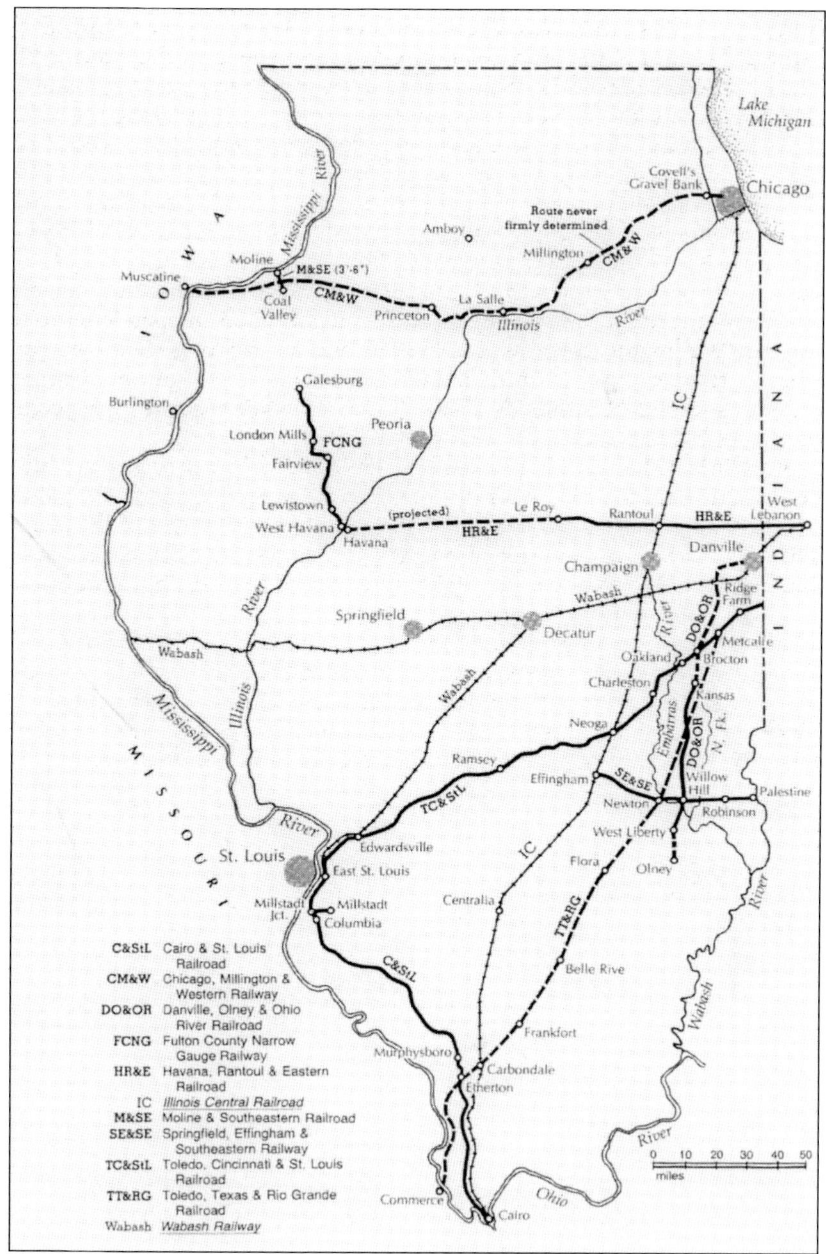

The conclusion of the Civil War in 1865 brought on an unprecedented era of railroad construction. The Union Pacific and Central Pacific completed the first transcontinental line in 1869, linking the nation together with a final golden spike. At the same time, small towns like Millington and Newark envisioned becoming cities if only they could gain enough rail traffic to bring in more businesses. One nearly forgotten historical footnote is the story of the Chicago, Millington & Western (CM&W), a narrow-gauge railroad. As shown on this map from George W. Hilton's *American Narrow Gauge Railroads*, the town of Millington would have been of utmost importance, because its location along silica sand deposits could assist in the development of glass-making businesses. Ultimately, the rails reached only as far as a gravel pit outside of Chicago. (Used by permission of Stanford University Press.)

This 1922 photograph of an engine bound for the newly built Bellevue & Cascade in Iowa permits a comparison of sizes between railroad gauges. A three-foot-gauge locomotive could easily fit on the bed of a standard-gauge flatcar that was four feet, eight and a half inches wide. Narrow-gauge railroads reached their heyday in the late 1800s and early 1900s, but the insurmountable problem of these lines was always the same—the locomotives and rolling stock could not be used on the vast majority of other railroads across the country, all of which were standard gauge. (Author's collection.)

The Fox River was up to its old tricks in 1954 when downtown Millington got flooded again. This time, the water was waist high in some spots, and for a while, people got around in rowboats or canoes. The railroad tracks were a quarter-mile in the distance, and although floodwater reached them, train traffic was not stopped. (Millington, Illinois, Historical Society Museum.)

A close-up look at what would affectionately become known as the Dinky reveals it to be a single gas-powered passenger car that could seat 30 passengers and carry sacks of mail and milk cans. It served towns like Millington along the Burlington's Fox River Line, offering an economical way of providing passenger service. Photographed at Aurora (its easternmost stop) in 1942, its front end was painted bright red with yellow trim for better visibility. (Little White Schoolhouse Museum, Oswego.)

The Dinky stopped for passengers at each town, first going westbound between Aurora and Streator, like in this 1942 view of it at Oswego, then returning on an eastbound route. In April 1943, less than a mile west of this point, the Dinky was involved in the one serious accident in its long history of service—a freight train hit the Dinky head on, and the Dinky's engineer, F.E. Bishop of Galesburg, was killed. (Little White Schoolhouse Museum, Oswego.)

In April 1943, a freight train left Montgomery, just below Aurora, and headed southwest along the Fox River Line. Its engineer had picked up his orders, which were to be carefully read and understood, since the signaling and radio dispatching of today did not exist at that time. Proceeding as if it were a normal day, the steamer's engineer did not read all of his orders, one of which said that he was to wait at Montgomery until the eastbound Dinky cleared the line. Tragically, the trains collided head-on near today's Boulder Hill subdivision. The engineer of the Dinky, F.E. Bishop, was killed, and several passengers were injured, but onboard was 16-year-old Helen Gilmore, who bravely led her fellow passengers to safety. In the ensuing fire, some sacks of mail were destroyed. It is not known if anybody from Millington or Newark was aboard the train. But this was not the end, as an identical Dinky was brought up from Galesburg to fill in until repairs could be made, then it continued to serve villages along the line until 1952. (Roger Matile, Little White Schoolhouse Museum, Oswego.)

The last run of the Dinky was on February 4, 1952, and among the photographs taken that day was a rather picturesque one of it crossing the bridge over the Fox River at Sheridan—a bridge that is still in use today. Interestingly, two decades ago, the idea was floated of running a dinner train over this same route, since it was quite beautiful, but the idea never got off the ground. As for the not-too-recent past, the reason for the discontinuation of this little train was the lack of riders, although people living nearby recall times around Christmas when an extra car had to be added because of an increase in passengers. Of course, that could not make up for the rest of the year, when the Dinky lost money. (Millington, Illinois, Historical Society Museum.)

The depot at Millington sat empty for years after the demise of the Dinky. Other small-town depots up and down the Fox River Line of the Burlington closed too. At Millington, the so-called house tracks (sidings used for spotting cars for loading and unloading) became storage tracks for outmoded Burlington equipment. Visible to the left of the depot is a retired coal tender from a steam locomotive that is waiting to be taken to a scrap yard. (Millington, Illinois, Historical Society Museum.)

Millington once had a grain elevator, and the Burlington Railroad served it. As with many wooden buildings from 100-plus years ago, fire damaged it, but the structure was rebuilt. One of the last times it caught fire was on June 20, 1929, when, according to the *Earlville Leader*, lightning struck and did $20,000 of damage; the loss of grain inside was valued at $2,000, with several nearby businesses also sustaining damage. (Millington, Illinois, Historical Society Museum.)

Farming was the main business of both Millington and Newark, and 100 years ago, it looked a lot different than it does today. Neighbors helped neighbors during busy times, as there was plenty of work to be done. Barn raisings and threshing bees were community activities. The names of the people and the location of the photograph have been lost to time, but what catches the eye is the steam-powered tractor in the background. A century ago, steam drove not only trains but also tractors. It was used to heat buildings in winter and powered factory machinery. A railroad was the way to the outside world, whether it was freight to be shipped or a ticket on a passenger train. The 1920s are regarded as the time with the highest number of miles of railroad trackage around the country—a golden age of sorts, although a decline would come soon after. (Fern Dell Museum.)

This c. 1895 picture of downtown Millington immediately tells the viewer that it was still the horse-and-buggy era—no automobiles are visible anywhere. The building on the right is the bank, which is still standing but today is a bar. Also of interest here is the boxcar sitting on the Burlington tracks leading to the depot and perhaps waiting to be loaded at the town's grain elevator. (Millington, Illinois, Historical Society Museum.)

Two decades after the previous picture was taken, much had changed in Millington along with the arrival of the automobile. One car is on the left, and more are visible down the street, although the horse-and-buggy era was still far from over. Electricity was available before the start of the 20th century, and around this time, telephones were being installed in homes and businesses. Near the far lower right is where the railroad tracks crossed Church Street. (Millington, Illinois, Historical Society Museum.)

The discovery of silica sand deposits along the Fox River brought a new industry to town in the late 1800s. The powdery white sand is used to make glass and cement (it is also now used in fracking for oil) and proved to be profitable to the Burlington, which hauled it. Soon, there were as many as 13 mines up and down the river. This image of the Millington Sand Pit is typical of what they looked like, with 20 men employed there in 1881. A fire in June 1892 spelled the beginning of the end, and the glass factory planned for Millington was never built. (Millington, Illinois, Historical Society Museum.)

Before the automobile, the makers of buggies had an important job, and almost every town had one. Newark had one (shown here), as did Millington. Buggies were how people took short trips from town to town in the 19th century, but if they wanted to go any real distance, they would have to take a stagecoach. This remained true until railroads came along and made long-distance travel possible for the masses. (Fern Dell Museum.)

Many Newark businesses are on display in this 1875 photograph. The shop on the right belonged to John Lawson, dealer, manufacturer, and repairer of boots and shoes. Next to Lawson was the Tin Shop, and on the left is what appears to be a barrel maker. At the time, bulk quantities of many items were shipped in barrels—one might contain salt, nails, pickles, apples, or almost anything. Because they could be easily rolled, barrels were commonly loaded onto freight cars for shipment by rail. (Fern Dell Museum.)

The business of welding was important because the metal tools used in farming often needed repair work, so in Newark, Dierzen's was the place to go. A little steam-powered tractor is on hand, although it is not known if it was in for repair work that day. The blacksmith shops of the past gave way to welding shops and automobile mechanics once the horse-and-buggy era ended. (Fern Dell Museum.)

If the blacksmith became the mechanic, was it possible for a mechanic to become an engineer? The answer to that might be Newark's Bill Thorsen, shown here at the throttle of No. 4 as it steams out of the Midland's train shed. Thorsen spoke of working once on cars when he was approached by men from the Newark Farmers Grain Elevator Company about doing some repair work. The subject of him running their train came up, and Thorsen said he would give it a try; it became his life's work. (Fern Dell Museum.)

This c. 1900 photograph features the then-new school simply named Newark School. This fine two-story brick building is where both elementary and high school students were taught. Two decades after the 1880 fire that had destroyed the Fowler Institute, a Christian-based private school with as many as 200 upper-grade pupils, a new public school would meet the town's need for higher education. Newark was ready for an influx of people who might come if a railroad was built through town, bringing with it new businesses and industry. (Fern Dell Museum.)

Newark's first pumper fire truck was photographed around 1927 in front of a farm implement store. The importance of such a piece of equipment cannot be overemphasized, since any fire outside the city limits (like the one that destroyed the railroad's bridges) would have to be fought by bringing in water. Hydrants did not exist outside of town, and nearby Millington had no hydrants or fire department and had to rely on Newark. (Fern Dell Museum.)

The importance of a railroad coming to town can be illustrated with a photograph of what was going on at the nearby town of Sandwich. The Burlington had built one of its main lines going west from Aurora during the Civil War, and Sandwich was one of the towns that benefitted from this. Visible in the background are industries that sprang up—something Newark and Millington hoped for but did not see. There was a double-track mainline and some sidings, while the Burlington's single track through Millington was essentially a branch line operation. Passenger service at Sandwich was different too. It was possible to take a train west across the entire state of Illinois, and eastbound trains went into Chicago. From the line through Millington, it was possible to go only as far southwest as Streator, and eastbound service ended in Aurora, where it would be necessary to buy another ticket and board a different train to reach Chicago. (Millington, Illinois, Historical Society Museum.)

Six

A Fiery Ending

How did vandals bring an end to the Midland, and what happened to lead up to that?

When newspaperman Grover Brinkman wrote his story titled "Illinois Farmers Claim Shortest Railroad" for the *Belvidere Daily Republican* on May 12, 1967, he had no way of knowing that the railroad would be gone by the end of the year. Fires set by vandals caused irreparable damage to two bridges located next to Millington, after which the board of directors at Newark Farmers Grain Elevator Company shut down the railroad forever.

Brinkman's story about the Midland recalled its earliest days, although Samuel Durant was not mentioned by name. Instead, the article concentrated on a concise history of the railroad since the end of the 1950s.

The article explained that as the 1950s drew to a close, the Midland was in desperate need of a little help; its locomotive was worn out, and its track needed repair work. The end seemed imminent. As luck would have it, the Illinois Agricultural Association got wind of the farm railroad and its woes, then decided it was within the association's realm to lend a hand. The case went all the way to Washington, DC, and with an assist from the Burlington, an agreement was reached so that the Midland would begin receiving a small portion of the freight charges for hauling grain from Newark's elevator to Chicago. The Burlington saw this as an opportunity to keep its business moving some 300 cars per year, while the tiny Midland's share would reap about $8,000 yearly, keeping it out of the red.

Tracks were repaired, a new locomotive was bought (secondhand, of course), and the future looked a little brighter. As Brinkman said, without the railroad, area farmers would have had to truck their grain 25 miles to Morris, then put it on a barge to Chicago's markets—a costly process freightwise. Brinkman continued his story, saying that if engineer Larry Akre had to blow his horn to chase cattle off the tracks, the Midland was still Newark's "bread and butter."

Fires set by vandals in October 1967 brought an end to the Midland.

What happened is still a matter of disinformation and disbelief. The story of fires being set to two wooden railroad bridges did not make front-page news, but short articles, like one in the *DeKalb Daily Chronicle*, told of how some kids were on a school holiday. The newspaper called them "boys with idle time" who set the fires. Although a few neighbors saw some smoke from the burning bridges, it was assumed someone was burning brush near the railroad tracks, so there was no apparent reason for alarm. The damage was done, the tracks were impassible, and the cost to repair the bridges was impossibly expensive. Dick West, the Midland's engineer at the time, saw the fire and recalls, "Somebody told me it was a good thing I didn't take the train all the way down there that day, or I'd have ended up in the creek!"

Unfortunately, the burned bridges were a disaster from which the Midland could not recover. Since it had always operated on a shoestring budget, repairing anything of this magnitude was futile, so in December 1967, the board of directors at the grain company (which owned the railroad) made the decision to close the Midland. Trucks would now handle all of the Newark Farmers Grain Elevator Company shipments, leaving the Midland to slip into history.

Property that had been leased as railroad right-of-way reverted back to its owners, and all that remained was to rip up the rails. What were said to be the last pieces of Midland track were removed in July 1970, although some are known to remain in place to this day; however, most of the old railroad ties were grabbed by landscapers. A storage place was chosen to stash the ripped-up rails, since there appeared to be no immediate use for them. The storage structure turned out to be a silage shed (a trench with a roof over it) in the middle of a field outside of Newark. There, the old rails laid undisturbed and known to just a few people for nearly 30 years.

In the late 1990s, the founders of Newark's Fern Dell Museum saw an opportunity to sell small pieces of vintage Midland rail as a fundraiser. It was an instant success, and the sawed-off sections sold quickly; two-inch paperweights went for $5, larger pieces went for $50, and $500 was soon raised. Then, in typical Midland hard-luck fashion, disaster struck when nearly all of the remaining pieces of historic rail were mysteriously stolen. A lifetime resident of Newark, Lowell Mathre, told the *Aurora Beacon-News* on November 6, 1997: "It's discouraging to have someone steal what a lot of volunteers are using to build something."

Perhaps the biggest disappointment was the realization that the stolen Midland tracks had probably been sold as scrap metal. Suspicion followed disappointment when it became obvious that the caper would have taken at least a couple of men and a large truck to pull off. Nobody was ever caught, much like the elusive bridge-burning vandals of 1967.

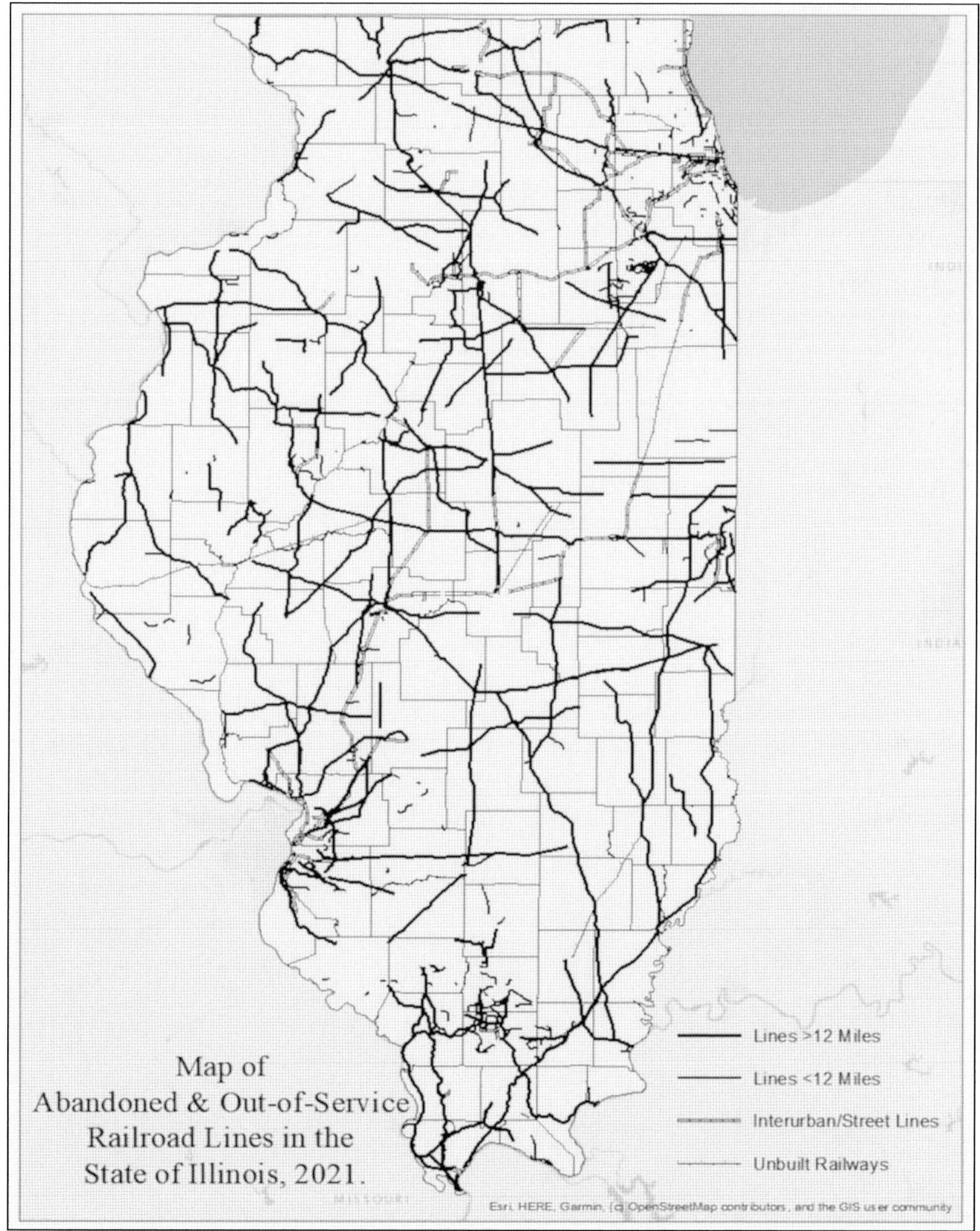

The rail scene in Illinois has changed a great deal since the time that the Illinois Midland Railway was built in 1913 and 1914, but had it reached its potential of a 120-mile route across the northern part of the state, its outcome might have been different—perhaps even successful. The plan to connect the cities of Rockford and Kankakee never did materialize, and no railroad completed such a route. On the drawing board, it had merit, since an outer beltline would have bypassed congested Chicago rail traffic. Instead, the Midland became known as the World's Shortest Railroad, which would require a magnifying glass to find on a map with today's other abandoned railroads in Illinois. Obviously, the state had too many miles of track that did not generate enough business to remain solvent. (Forgotten Railways, Roads & Places [ffandp.com].)

There seem to be no photographs that were taken while fire destroyed two Midland bridges, ultimately bringing the railway to an inglorious ending. Witnesses say the smoke they saw was simply thought to be from someone burning brush near the tracks on that October 1967 afternoon. Lowell Mathre saw it, as did Dick West, engineer on the train headed in that direction, who promptly stopped and drove it back to Newark. Lifetime Millington resident Doug Holley also witnessed the fires and says the two bridges that were burned were located closest to Millington. The resulting irreparable damage doomed the Midland, then the Newark Farmers Grain Elevator Company (owner of the Midland) switched to trucks for its transportation needs. One of the bridges targeted could be the one shown here, where the railroad had drawn water for its locomotives out of Clear Creek. If someone wanted to hurt the Midland, destroying that particular bridge (out of the five) might have been the way. Still, by 1967, the Midland had long been a diesel-powered railroad, and water was no longer needed for the boiler. Why would the vandals bother destroying a second bridge as well? (Fern Dell Museum.)

A view from the engineer's seat shows the Midland's main line trackage overgrown with weeds, vines, trees, and just about anything else that could grow in northern Illinois. This photograph also helps explain why the vandals setting the fires were not seen—they had plenty of cover. Were they "boys with idle time," as the *DeKalb Daily Chronicle* wrote, or was there more to it? Nobody was ever caught. That December, at a meeting of the board of directors of the Newark Farmers Grain Elevator Company, the decision was made to shut down the railroad, and from then on, trucks would handle all the transport of grain. A bit later, a rumor went around that the railroad had been put up for sale. Needless to say, there were no bidders (none were expected), and removal of the rails began. (Kendall County Historical Society.)

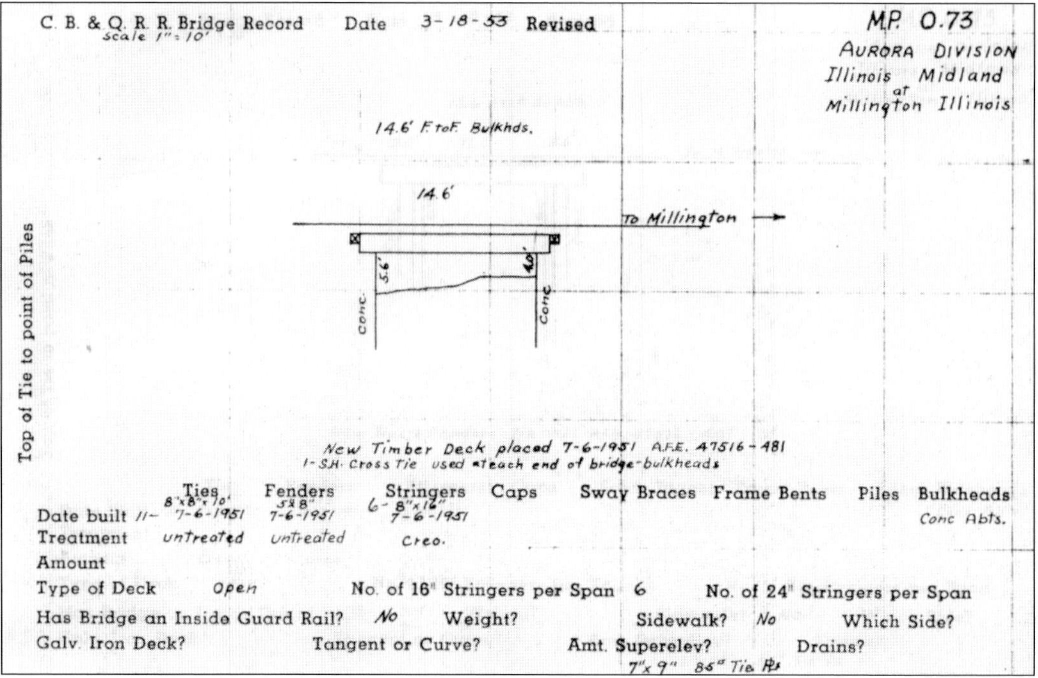

C. B. & Q. R. R. Bridge Record Date 3-18-53 Revised
Scale 1" = 10'

MP 0.73
Aurora Division
Illinois Midland at Millington Illinois

14.6' F to F Bulkhds.

14.6'

To Millington →

5.6' 4.0' Conc. Conc.

Top of Tie to point of Piles

New Timber Deck placed 7-6-1951 A.F.E. 47516-481
1-S.H. Cross Tie used at each end of bridge-bulkheads

	Ties	Fenders	Stringers	Caps	Sway Braces	Frame Bents	Piles	Bulkheads
	8"x8"x10'	5x8"	6-8"x16"					
Date built 11-	7-6-1951	7-6-1951	7-6-1951					Conc Abts.
Treatment	untreated	untreated	Creo.					
Amount								

Type of Deck Open No. of 16" Stringers per Span 6 No. of 24" Stringers per Span
Has Bridge an Inside Guard Rail? No Weight? Sidewalk? No Which Side?
Galv. Iron Deck? Tangent or Curve? Amt. Superelev? Drains?
7"x 9" 8½' Tie Pls

This blueprint for the bridge shown in a previous photograph may be one of the two located closest to Millington destroyed by fire in October 1967. This drawing explains that a new wooden deck was installed on July 6, 1951. The Midland contracted out such work to the Burlington and its buildings and bridges department. Most major railroads had crews for those tasks—but not the Midland. In many ways, the Burlington was like a big brother to the Midland. (Burlington Route Historical Society.)

Following the fires that destroyed two bridges and ultimately put the Midland out of business, the only task that remained was to pull up the tracks. This happened in July 1970. Shown at work on the dismantling of the Midland are (from left to right) Don Johnson, George Cook, Dave Johnson, and Chet Ramsey. They did not get all of it, and pieces of it remain in place to this day. (Millington, Illinois, Historical Society Museum.)

Of the few businesses served by the World's Shortest Railroad throughout its 53 years, none was more successful than the Newark Farmers Grain Elevator Company, which dated to 1915. Described by some as a sort of community enterprise, it remains an example of how people can band together for the common good of everyone. The company leased the railroad at first then bought it outright in 1942. (Fern Dell Museum.)

By the start of the 21st century, it was clear that larger, more modern facilities were needed for shipping grain from Newark. The Midland was long gone; trucks had taken its place, and with massive new storage bins in place, it was time to call in the wrecking crew. In 2008, the remains of the original wooden structures were demolished and hauled away. (Fern Dell Museum.)

At this point, it is a good idea to insert a modern picture of the current grain company in Newark, which is now part of CHS Grain, with facilities in 16 states and headquarters in Inver Grove, Minnesota. CHS has expanded into other areas besides agriculture but is still run as a co-op, and grain is still shipped by rail at some of its facilities. CHS stands for Cenex Harvest States Cooperative, and the co-op advertises by stating, "We help farmers grow profitable crops with targeted fertilizer and crop protection solutions." The company's third-quarter net income for 2021, as this book was being written, was reported to be $273.6 million. (Author's photograph.)

Seven

Midland Memories

Evidence of the Illinois Midland Railway—as well as remnants of other pieces of the past—still exists, but where can it be found?

Time passed and combined with natural disasters like fires and floods, and there were businesses that folded up or changed hands. Millington and Newark are still small but alive and well today. The success or failure of their railroad was not the whole story, but people who remember it can tell some interesting tales!

One incident from the early 1960s has become oft-told Midland lore guaranteed to get heads shaking in disbelief. It began on a day as ordinary as any other on the little railroad, with engineer Larry Akre at the throttle of the well-worn engine No. 9120 (if renumbered, would have been Midland No. 5) switching cars around Newark's small yard. For some reason, he had a helper with him that day—a brakeman who was assisting Akre in getting cars onto the right tracks for loading or perhaps for the upcoming trip down to Millington. It is possible they were poling a boxcar, which involved shoving a car a short distance without having to couple or uncouple it; this was a simple but risky practice.

Exactly what went wrong is not clear, but for some reason, the brakeman hopped aboard the moving boxcar, climbed up the rear ladder to reach its brake wheel, then attempted to tighten it, bringing the car to a stop. But it did not stop! The malfunctioning hand brake turned freely like a bicycle wheel, so the car kept rolling faster, headed downhill. Now fearing for his safety, the brakeman bailed, leaving the out-of-control car rolling down the tracks to Millington with nothing to stop it. By coincidence, Lowell Mathre and a friend happened to be standing trackside when they heard the familiar clickety-clack of an oncoming train, but there was no train. They reeled back, staring as the runaway boxcar went zipping by and onto the Burlington's main line, headed somewhere west. A railroad disaster was almost certainly in the making, yet there was nothing anyone could do but watch the car roll out of sight.

Fortunately, the errant car finally harmlessly came to a stop halfway to Sheridan (about two miles), and amazingly, nobody got killed. The Midland crew pursued it and rescued the car, so no harm was done. Railroading can certainly be a dangerous job.

As for that elusive Midland track, it is related to yet another hard-luck story that involves Jimmy Judd in the early 1980s, shortly after he had moved to Millington and was delighted to discover a small portion of what seemed to be remaining Midland track on his property. Having heard about the fabled World's Shortest Railroad, he asked the powers that be in Millington if he might be allowed to place some sort of historical marker next to it, adding that he would pay all the costs involved. Not only was he told no, he was ordered to remove the track, and when he refused, the village did it for him. It seems that not everyone in town had a fondness for old railroad stuff.

Finally, there is an iconic piece of local lore that needs to be told. Although its relation to the railroad amounts to very little, it borders on the bizarre. This occurred a couple of generations ago, so some of the details have been lost to time.

A nice family lived in a small house that was probably owned by the grain company, which owned the railroad and just about everything else nearby, and was located next to the Midland's engine shed. Photographs exist of the house, including one on page 35 of this book that shows a snowy winter scene taken from atop the Newark grain elevator with the house pictured just west of the shed. Nothing was unusual about any of this; not much was known about the family who lived there, including what the father or mother did for a living or who their kids were and where they went to school, but it all seemed relatively normal.

Then came the morning when nobody from the house showed up where they were supposed to be, and something seemed amiss. When a few townspeople went to the house to investigate, they were astonished to find nobody at home. Inside, bowls of breakfast cereal sat on the table, the radio was tuned in and playing, clothes hung in the closets, furniture was still in place, but there was no sign of the people who lived there. It was as though they had vanished from the face of the earth, because the family was never seen or heard from again!

A house with such a stigma in a small town subsequently sat for years and was used to store lumber before it was ultimately demolished. Meanwhile, it had garnered a nickname and was known around town as the spook house.

Through the years, so many Illinois railfans have said they grabbed a cab ride or two on the Midland that an all-time roster of folks who rode the train might be longer than the list of those who missed the chance.

On display in City Park on Front Street in Newark is the last remaining piece of Midland rolling stock. It is a small handcar trailer car that can be spotted in photographs from the 1940s, when it was attached to the Model T track inspection vehicle. Somehow, it got saved after the railroad shut down at the end of 1967, and for the 1976 Bicentennial, it was painted, put on a piece of original track, and had a plaque attached to it. The plaque reads: "Illinois and Midland Railroad / Shortest Railroad in the United States / Sponsored by Kendall County Council of the American Legion Dept. of Illinois / This Marker Placed by the Kendall County Bicentennial Commission 1976 / XVK Foundry Millington Illinois." (Author's photograph.)

A closer look at the Midland track in City Park reveals an interesting story. It did not take much to scrounge up some original track, since it did not all get torn up or disappear, although someone with a hacksaw must have cut out this piece. Both local museums—Fern Dell and that of the Millington, Illinois, Historical Society—have nice examples of track on display, but lightweight rail proved to be problematic for the railroad. Rail data charts place this size of rail at 60 pounds per yard or less, which is too light to carry any heavily loaded cars. Some original ties were located for the display in the park, although most of them had been pulled out by a couple of men who used them in their landscaping business. All things considered, a display of something that was once part of the Midland has survived for anyone to see. (Author's photograph.)

The Fern Dell Museum, so named for a schoolhouse that once bore the name, is located directly across from City Park on Front Street and is the largest repository of Illinois Midland Railway memorabilia that can be viewed and touched. Started in 1995, the museum been located at its present quarters since 2009. Previously, the building housed the Leland Hatchery Store (a good business in a farming community) as well as an antiques shop. (Author's photograph.)

On the same side of Front Street as the park where Midland's handcar is on display is the bank where Samuel Durant stashed the money that was to be used in building the Midland, most of which ended up in his pockets. What was once the Farmers State Bank of Newark still stands after more than 100 years, but it is no longer used as a bank. It now houses a business that sells and repairs golf carts. (Author's photograph.)

This photograph dates to the 1940s, when the surviving handcar trailer was still in service on the Midland. At that time, the quaint track repair operation, using a Model T, was guaranteed to turn heads. The automobile wore out and was sent to a junkyard. Still, the continued existence of something from those days is what nostalgia is all about. (Fern Dell Museum.)

When it comes to Illinois Midland memorabilia, there are a variety of places to look, and sometimes it will turn up where least expected. Here, what is said to be the last piece of Midland track in Millington is still sitting in its original location, although it is partially buried in an undisclosed front yard. Railfans are reminded to respect the rights of private property and ask for permission before entering anyone's yard. (Author's photograph.)

Inside the Fern Dell Museum is a model railroad in HO scale that helps visitors envision how the Illinois Midland once looked. The tall building near the center is a model of the grain elevator, while the track configuration represents how the small yard was designed and built dating back to 1915. The yard is where freight cars were switched in and out and moved into place for loading in a very efficient operation. To the left of the grain elevator's peak is a model of the enginehouse complete with an HO locomotive resembling a saddle tanker, and just beyond the enginehouse sits a small two-story home that is supposed to represent the spook house (the story of this house is told in this chapter's introduction). The left side of the photograph has a preserved newspaper page of the *Chicago Times* article that reporter Judith Allen wrote in 1946 when she visited Newark. Ten years later, the *Chicago Tribune* ran a very similar human-interest feature, but by then, the Midland's engineer was Mel Severson, and the general manager of the grain elevator and thereby the railroad was Gene Kunkel. Credit goes to Bruce Moore and Lynn Perlke for building the layout. (Author's photograph.)

Fern Dell has a display of Midland track that helps people to understand just how lightweight it was, raising the question—how did this hold up a 40-ton locomotive or a boxcar loaded with grain? The lightweight track was cheaper but in the long run proved to be detrimental to the operation of the Midland. Trains could not go much faster than three or four miles per hour without risking derailment, and the track needed constant maintenance. On the left side is a piece of modern 131-pound-per-yard rail that was placed there for comparison. (Author's photograph.)

Samuel Durant had purchased the secondhand rail to be used in building the Midland from unknown sources. Where some of the rail was produced was clear, since the name of the steel mill and the date of production had been cast into the sides of some of the sections. This example was made in Joliet, Illinois, in 1888, so it was produced—and likely in use—25 years before coming to the Midland. (Author's photograph.)

Lowell Mathre, a lifelong resident of Newark and its former fire chief, is shown admiring the bell and whistle for Midland locomotive No. 4, part of the collection at the Fern Dell Museum. No. 4 was the last steam engine on the railroad and was withdrawn from service in 1959. Attempts to preserve and display it in City Park failed, so it was sold to a scrap yard dealer. As luck would have it, the scrap yard dealer was not making much money on his investment and realized the historic value some of the parts might have, so he gave the folks in Newark a call. He asked: Would they be interested in buying the bell, whistle, and a couple of other parts—for one dollar? Mathre says they jumped at the chance, and once everything came back to Newark, former engineer Mel Severson was on hand; when he heard them ring the bell, it brought tears to his eyes. (Author's photograph.)

In this picture taken at Fern Dell Museum, Michael Kehoe (left) listens as Lowell Mathre explains the history of Newark's Fowler Institute. Started in 1855, it was designed to meet the town's need for higher education—meaning more than the three Rs that were being taught in one-room schoolhouses at the time. Fowler Institute succeeded, with yearly enrollment of some 200 students, and Newark's citizens were proud to have an academy; then, they thought, if only they could get a railroad. When the Ottawa, Oswego & Fox River Valley (OO&FRV) Railroad was built in 1871, it went through Millington instead of Newark. The Fowler Institute burned down in 1880. (Author's photograph.)

What is the oldest building in town? Without hesitation, Lowell Mathre answered, "the blacksmith shop." In a fine example of restoration, the 1840s blacksmith shop is open on special occasions, and demonstrations given there help bring history to life. Fletcher Misner, who built his first blacksmith shop in Millington before 1840, also had a hand in building this one in Newark. (Author's photograph.)

Clear Creek still flows from Newark to Millington, and it is still overgrown—like it seems to have always been. This is where the Midland stopped to take on water from the creek for its locomotives, where simple wooden bridges were built for trains to get across it, and where vandals set fire to two bridges in 1967, which resulted in the end of the Midland. As a reminder to railfans who might want to explore the area, this is now private property. One story concerning the creek that continues to receive attention is that there is quicksand somewhere down there. However, lifetime resident Jeff Mathre dismisses this story as a myth. Although the ground is soft and soggy from its high water table, no quicksand exists along the banks of Clear Creek. (Author's photograph.)

When the question "What's the oldest building in town?" is asked in Millington, the answer sounds like an echo: it is the blacksmith shop. Fletcher Misner built Millington's blacksmith shop along the Fox River around 1839 using sturdy limestone, but it was not his first try. The first attempt, made of logs and wood, was washed away in a flood. Blacksmiths were courted by towns that needed them, much like towns that were trying to get a railroad built nearby. An enterprising fellow, Misner also constructed and sold buggies. By the start of the 20th century, his shop had become Fred Leonard's button factory, then later the building was used as a foundry under a variety of owners. An interesting footnote to the building being used as a foundry is that it is where the plaques for the nation's bicentennial were cast before being placed around Kendall County in 1976. (Author's photograph.)

Built in 1922, the Millington School had two classrooms and served the community until 1975, when declining enrollment forced its closure. After that, students attended different schools in two nearby districts because Millington, due to a geopolitical quirk, was divided into LaSalle and Kendall Counties, so the town was in two separate school districts. This building sat empty and was occasionally used for town hall meetings before 2017, when half of it was taken over by the newly formed Millington, Illinois, Historical Society. (Author's photograph.)

At the Millington, Illinois, Historical Society Museum, director Beverly Casey (right) is explaining to Jenn Kehoe the earliest history of the town, which dates to the time Indigenous peoples inhabited the land along the Fox River. A fine collection of artifacts from those precolonial times is on display, with many pieces donated by local residents. The Fox River was a source of food for Native Americans; Chief Blackhawk became their legendary leader, and an effort by members of the Sauk and Fox tribes to reclaim their land became a factor that helped spur the Blackhawk War of 1832. Two years after the Indigenous peoples were driven out of the area, new residents began to arrive in Millington. As for railroads, the first one that came through the area—the Ottawa, Oswego & Fox River Valley—arrived in 1871. The Midland was not built until 1913, and the museum is always looking for more railroad photographs and memorabilia to add to its growing collection. (Author's photograph.)

Janet Luther Blue, a lifetime resident of Millington, shows how a clamshell could be used to make buttons; a half dozen or more could be drilled from a single shell. Millington's button factory, which was located in the oldest building in town (it once was the blacksmith shop), prospered for a time; buttons could be shipped as small freight packages from the railroad depot, and examples of buttons made in Millington are on display in the Millington, Illinois, Historical Society Museum. The advent of plastic buttons spelled the end for using clamshells. Blue also spoke of how her father trapped muskrats and turtles along the Fox River many years ago and that he sold the hides and meat for a little bit of extra income for the family. (Author's photograph.)

People who remember the Midland know that it never owned a caboose, but they saw a lot of this one. Beautifully restored inside and out, this Burlington cupola-type caboose (a "Q" waycar, in railroad parlance) has been on display since 1974 at Lyon Historic Farm in Yorkville. Built in 1918 by the Burlington's Aurora Shops, it was in service for 56 years on the Fox River Line and would have been seen almost daily at Millington, where freight cars were interchanged between the Burlington and the Midland. Cordial relations were maintained between the two railroads for many years. (Author's photograph.)

Steam ruled the rails for more than 100 years, but many people may not realize it was also used to power farm equipment. Steam tractors, like the one shown here at Lyon Historic Farm in Yorkville, once plowed fields across the Prairie State of Illinois. That kind of advancement certainly made farming easier, since it beat using a team of horses or mules to do the job. The principle behind its operation was the same as with a steam locomotive—the boiler had to be fired up to create a head of steam, with the pressure from the steam causing pistons to move and wheels to turn. Note the cleats on the wheels to provide traction and prevent it from getting stuck in the mud. This antique tractor is fired up on occasion and is quite a sight to behold. (Author's photograph.)

The Burlington's Fox River Line, which dates to 1871, is alive and well, as shown in this photograph taken in 1990 at Yorkville. The end of the Midland, with its connection to the Burlington and the loss of the little bit of business it brought in over the years, was hardly felt. By 1990, silica sand had become the line's major cargo, and soon, it was the only cargo. Here, Burlington Northern's No. 11841 is headed toward Wedron, the location of the major silica sand mine that has been in business for over a century. Also of interest is the grain elevator along the weed-choked siding visible on the left side of the photograph. Like in Newark, rail service to the elevator ended years ago, and now trucks transport the grain to market. (Author's photograph.)

The Fox River Line was a bit of a dusty corner in the Burlington Northern's vast rail empire. It made money hauling silica sand, which is what is in the covered hopper cars in this picture from March 1992. Entire trains filled with the sand went up and down the line, usually twice a day, with a caboose on the end. That was an unusual sight in itself, because railroads had retired nearly all of their cabooses a decade earlier, but this one was probably still in use as a backup car at Wedron. A crewman with a radio in the caboose could safely guide the engineer as he backed a long train onto a siding. The Fox River Line was sold to Illinois RailNet, which runs the same operation today minus the caboose. (Author's photograph.)

Aurora, Illinois, is recognized as the birthplace of the Burlington Railroad going back to the 1850s, when railroads were just beginning to take hold in the United States. In the Burlington's shops, workers built the locomotives and rolling stock that would be used to provide freight and passenger service to the towns in the area. An important Burlington main line extended into Chicago, the city that would soon become the rail capital of the country. It is natural that towns like Millington and Newark wanted to get a railroad of their own, and Aurora was barely 20 miles away, so the Burlington looked like a good prospect. A lot has changed over the years, but part of the Burlington's shops are now beautifully restored and used as a commuter station as well as a brew pub. To honor the men and women who once worked there, a statue was erected in front of the station. (Author's photograph.)

One way to discover the ages of Newark and Millington might involve a trip to the cemetery they share, the Millington-Newark Cemetery on Fox Ridge Road. Once inside, it does not take much looking to find a grave that is usually decorated and well cared for—that of Henry Misner (1760–1848). He fought in the Revolutionary War, and it is interesting to think that someone buried in that cemetery was around before the United States became a country. In 1897, the Daughters of the American Revolution placed a newer, bigger marker on his grave. When Samuel Durant and his Illinois Midland proposed building a railroad from Millington to Newark, they had to go around this sacred ground, which was accomplished by leasing strips of property from local farmers instead. (Author's photograph.)

A look around the Millington-Newark Cemetery leads to more interesting history. The graves of Nat Jones (1826–1904) and his wife, Celeste (1836–1900), are there. Nat Jones was a formerly enslaved person who served in the Union army during the Civil War and later got married and settled in Newark, where he farmed. No railroad was going to be built through the cemetery, so the Midland went around it instead. If one follows the service road to the end of cemetery property, it is possible to see where the Midland's track once passed by looking east along a tree line about 100 yards across the field. (Fern Dell Museum.)

The story of the Illinois Midland Railway is also a celebration of steam, which is part of why it is so attractive to railfans. It was steam that powered the railroads of a bygone era that are now part of the fabric of our history. Today, is there anywhere a person might still see a steam engine chugging down the tracks? The answer is yes—at a railroad museum. Most railroad museums have examples of steam locomotives, but they are usually just sitting outside gathering rust. Finding one that is still running is hard, but it can be done. Not too far from where the Midland once ran, just 56 miles away, is the Illinois Railway Museum in Union, Illinois. On special weekends, steam locomotives are fired up and put on tracks, and it might even be possible to grab a picture like this. As the older gentleman is leading the little boy by the hand to get a good look at that engine, he might be saying, "Now, back in my day . . . " (Author's photograph.)

Consistent with our mission to preserve history on a local level, this book was printed in South Carolina on American-made paper and manufactured entirely in the United States. Products carrying the accredited Forest Stewardship Council (FSC) label are printed on 100 percent FSC-certified paper.